Gated Communities

This informative volume gathers contemporary accounts of the growth, influences on, and impacts of so-called gated communities, developments with walls, gates, guards and other forms of surveillance.

Gated communities have become a common feature of the urban landscape in South Africa, Latin and North America. It is also clear that there is now significant interest in gated living in the European and East Asian urban context. The chapters in this book investigate a range of social, political and urban issues thrown up by the emerging proliferation of fortified styles of community life. These include:

- Emerging trends in gated communities around their governance and proliferation
- Gated communities in developing countries, including:
  - the wider social impacts of gating and 'forting-up'
  - the links between gating, segregation and urban form

The chapters in this volume enable the reader to understand more about the social and economic forces that have lead to gating, the ways in which gated communities are managed, and their wider effects on both residents and those living outside the gates.

This book was previously published as a special issue of the journal *Housing Studies*.

Rowland Atkinson is a senior researcher at the School of Sociology, University of Tasmania, Australia.

Sarah Blandy is a lecturer at the School of Law, Leeds University, UK.

Gated Communities

Edited by
Rowland Atkinson and
Sarah Blandy

LONDON AND NEW YORK

First published 2006 by Routledge
2 Park Square, Milton Park, Abingdon, Oxon, OX14 4RN

Simultaneously published in the USA and Canada
by Routledge
270 Madison Ave, New York NY 10016

Routledge is an imprint of the Taylor & Francis Group

Transferred to Digital Printing 2008

Typeset in Times 10/12pt by the Alden Group Oxford

British Library Cataloguing in Publication Data
A catalogue record for this book is available from the British Library

Library of Congress Cataloging in Publication Data
A catalog record for this book has been requested

ISBN10: 0-415-37315-8 (hbk)
ISBN10: 0-415-46379-3 (pbk)

ISBN13: 978-0-415-37315-9 (hbk)
ISBN13: 978-0-415-46379-9 (pbk)

CONTENTS

Introduction: International Perspectives on The New Enclavism and the Rise of Gated Communities

ROWLAND ATKINSON & SARAH BLANDY

This volume gathers together substantially revised versions of papers first presented in Glasgow, Scotland in September 2003. This event was an international meeting of academics who discussed the significance, relative problems and benefits associated with the international rise of gated communities. Gated communities (hereafter GCs) have been defined in a number of ways. These definitions tend to cluster around housing development that restricts public access, usually through the use of gates, booms, walls and fences. These residential areas may also employ security staff or CCTV systems to monitor access. In addition, GCs may include a variety of services such as shops or leisure facilities. The growth of such private spaces has provoked passionate discussion about why, where and how these developments have arisen. This volume presents an opportunity to gather together contemporary and diverse views on what is at least commonly agreed to be a radical urban form.

The apparently 'unique' characteristics of GCs present immediate problems for an accurate definition. Should we include flats with door entry systems, tower blocks with concierge schemes or partially walled housing estates, even detached houses with their own gates? Among this confusion we suggest that the central feature of GCs is the social and legal frameworks which form the constitutional conditions under which residents subscribe to access and occupation of these developments, in combination with the physical features which make them so conspicuous.

Living in a gated community means signing up to a legal framework which allows the extraction of monies to help pay for maintenance of common-buildings, common services, such as rubbish collection, and other revenue costs such as paying staff to clean or secure the neighbourhood. However, such legal frameworks can also be found in many thousands of non-gated homeowner associations in the US, and indeed in blocks of leasehold flats in England. This leads us back to the important physical aspects of these developments. Where a combination is found of these socio-legal agreements and a physical structure which includes gates and walls enclosing space otherwise expected to be publicly accessible, we can finally achieve some clarity of definition. Gated communities may therefore be defined as walled or fenced housing developments, to which public access is restricted, characterised by legal agreements which tie the residents to a common code of conduct and (usually) collective responsibility for management.

While this definition may be useful it is often argued that gated communities express more than a simple constellation of particular physical and socio-legal characteristics. In the built environment around us we increasingly see examples of an attempt to boost defensible space and the means to exclude the unwanted. This has meant that we can now observe a continuum of 'gating' which can be seen moving between symbolic and more concrete examples. Suburban areas with booms across private roads, housing estates with 'buffer zones' of grass and derelict land, and cul-de-sacs all express a mark of exclusion to non-residents with varying degrees of efficacy. All of these built forms suggest a lack of 'permeability' in the built environment directed at achieving increasingly privatised lifestyles, predominantly through the pursuit of security. It is this attempt at self-imposed exclusion from the wider neighbourhood, as well as the exclusion of others from the gated community, which has driven a much wider debate about the relative merits of gating and other strategies to achieve security, when set alongside other key concerns such as freedom of access to the wider city, social inclusion and territorial justice.

These various issues form a core concern for analysts of gated communities—are they really just a different type of housing or are they an extension of a rationale attempting to create control, predictability and personal safety which may have external and negative secondary impacts? Should gated communities be condemned or their example extended, as the current Home Secretary has recommended. Regardless of these various questions it is clear that gated communities have already divided many observers, critically over whether ability to pay for security and privacy should allow such voluntary social exclusion and access to these bubbles of safety. Indeed, the need for GCs may be challenged in the context of the UK where prevailing crime rates may not seem to warrant fortress-style enclaves. As we will see, these questions are by no means easy to address, either for academics or policy-makers seeking socially equitable development.

The Fortified Neighbourhood

It is now well documented that gated communities can be seen as a response to the fear of crime (Atkinson *et al.*, 2004) but other drivers also appear significant. In particular the desire for status, privacy and the investment potential of gated dwellings all form important aspects of the motivation to live behind gates. For many housing researchers drawn to new social problems and forms, gated communities appear a profoundly interesting and relatively new object of study in the European context. The privatisation of public spaces, fortification of urban and residential space and the embodiment of public fears about private crimes (both property and personal) that gated communities invoke provide rich ground for urban and housing researchers.

While commentators in Latin America, the US and South Africa have witnessed a massive upswing in gated development for some time, their impacts and costs have only begun to be speculated upon in the European context. In the British context little has been written on gated communities, though this is changing (Webster, 2001). Recent consideration of the value of gated communities in the UK has been challenged by planners who view them as exclusive, unnecessary and burdensome due to the restrictions on movement that they promote (Minton, 2002). A recent study suggests that there are now more than 1000 gated communities in England alone (Atkinson *et al.*, 2004). While this may be numerically insignificant, the wider symbolic character of such development at a time when the government is committed to pursuing styles of development which allow permeability, affordable housing and diverse housing tenures appears problematic. There seems much to learn from an international perspective on such issues.

Outside the European context gated communities have had a much longer genesis; as a built form, gated enclaves were common in many areas of the world. In America they remained a relative rarity until large master-planned communities arrived in the 1960s (Blakely & Snyder, 1999; McKenzie, 1994). Many have argued that GCs represent a search for community with residents seeking contact with like-minded people who socially mirror their own aspirations. While advertising by developers (primarily in America) draws on this communitarian ideology it has been clear to some that the idea of a gated 'community' is something of an oxymoron. Increasing numbers of recorded neighbour disputes and conflict between residents and their management companies suggest at least as many problems as are found in 'normal' developments (see for example, Linford, 2001). An American survey carried out in 1987 found that 41 per cent of associations suffered from major problems with rule violations (Barton & Silverman, 1987). In this volume Evan McKenzie picks up on this theme and argues that gated communities increasingly contain residents openly hostile to the strictures to which they have signed up, or faced with potentially massive problems of affordability due to large repair bills for common roads or facilities. The possibility that GCs contain some kind of built-in obsolescence may become increasingly apparent.

The Inter-Neighbourhood Contract: Place, Social Justice and the Pursuit of Privatism

A key driver of interest in GCs stems from the sense that they form an intellectual intersection at which we can locate a much wider range of social changes and concerns relating to our urban context. The existing research agenda has focused on the residential choices of a select demographic group, largely characterised by self-interest and personal affluence as well as a desire for disengagement. However, this is to miss much of the point at stake in the analysis of gating. Gated communities express a broader trend of private decision-making that has wider and public ramifications. In short, the locational choices made by affluent households affect outcomes for the poor in terms of city sustainability, security and social segregation. If 'forting up' is taken to extremes this search for security will have enormous impacts on those left outside these new enclaves. Our theme in this introduction is that the choices of these relatively few gated dwellers may help us to conceptualise what might be thought of as a kind of spatial contract which, if not balanced by public intervention, may lead to a downward spiral of urban social relations. Let us expand on what we mean by this.

Many observers have suggested that urban segregation has represented the crystallisation of wider social divisions and problems that are largely negative in their impact. For earlier writers, like Gans (1968), the importance of socially diverse areas lay in the empathy generated by meeting people of different social backgrounds and experiences. Increased concentrations of poverty and clustering along ethnic and socio-economic lines has left many cities divided in ways that commentators believe hinders political empathy while concentrating disadvantage and exclusion from employment and educational opportunity (Massey & Denton, 1993). To read more widely from this, the residential choices of society at large have important secondary impacts on those with least choice and whose concentration dislocates and disconnects them from prospects for personal development (Atkinson & Kintrea, 2002). There are, then, reasons to believe that segregation is problematic.

The process of gating surrounds an attempt, in part, to disengage with wider urban problems and responsibilities, both fiscal and social, in order to create a 'weightless' experience of the urban environment with elite fractions seamlessly moving between secure residential, workplace, education and leisure destinations (Atkinson & Flint, 2004; Graham and Marvin, 2001). However, this apparent floating world of the rich is still connected to the lives of those living in other areas through the tendril linkages of taxation, legal contract and a system of social policy interventions which attempts to bridge and ameliorate these social divides. Centralised taxation and spending systems cut across place of residence and link people of diverse social positions, this much is perhaps self-evident but is clearly significant.

Increasing ghettoisation is occurring in a bifurcated manner with groups at both the top and bottom increasingly concentrated together in socially homogeneous areas. These processes are linked by fiscal and social contracts but which are now threatened by calls for at least partial fiscal autonomy by gated communities and the social withdrawal of the affluent. As these groups pull out, the effects of social concentration and residualisation further down a hierarchy of neighbourhood desirability are increased. Another way of thinking critically about this new enclavism is to consider why ghettoisation of the poor has so regularly been considered problematic while affluent concentration is not. The answer by those predisposed toward market solutions is likely to be that compounded problems of social disadvantage formed by concentrated deprivation are not to be found in areas of concentrated wealth which are therefore the 'answer'. However, as we contend here, these processes are linked and mediated by the local and central state and housing market in an increasingly inequitable way.

Gated communities appear as segregated spaces with a social ecology that is planted into the fabric of the city; where the wall starts a new social area begins, whether one lives inside or out. A key question remains, if this ghettoisation of the affluent proceeds how will this affect the ameliorative social ties negotiated through the state—of welfare, social services and of crime control for those living in ghettoised poverty? While the club system of private access to security (Hope, 2000) allows the affluent to displace crime this access to safety is denied those with fewer resources.

So far the debate on GCs has largely been considered in terms of whether the freedom to make locational decisions that maximise personal security should be the overriding feature of public policy debates about the relative value of such developments. Linking perhaps legitimate fears to personal residential choice is difficult to challenge in the abstract and yet, in many areas of social life where the decision to go it alone would have deleterious consequences for others, the state regularly and legitimately intervenes to reduce negative wider outcomes. To act as personal law-maker, to build where we like or to forego taxes are actions intervened in by institutions of governance in order that wider impacts are prevented. Gated communities represent a desire for accentuated positive freedoms (the ability 'to do' something) but which hinder the negative freedoms (Berlin, 1958) of others (the ability to be 'free from' something, such as increased crime displaced by the presence of a GC) in adjacent neighbourhoods.

Even before getting into a debate about the relative merits of gating we find systematic research which suggests that the shelter from fear that gated communities appear to represent soon fades once residents move in. Research by Low (2003) suggests that living 'behind the gates' actually promotes fear of the unknown quantities of social contact outside them. The lack of predictability and experience of people in social situations outside these compounds appears to play out most strongly for the young, particularly those brought up in gated communities.

In fact, perceived safety and actual crime rates have been found to be no different between gated communities and similar, but non-gated, high-income American neighbourhoods. This research suggested that a sense of community was higher in the non-gated neighbourhoods (Wilson Doenges, 2000; confirmed by Sanchez *et al.*, forthcoming). These various issues lead us into a whole new area: the long-run consequences of creating enclave-style developments. What happens to people who live in areas where social distinction is also expressed through clear physical boundaries (between a socially homogeneous affluent group and the mass beyond the gates)?

The impacts of withdrawal by select social groups has wider implications for how we evaluate social justice in the city context. We suggest that the internal contract of GCs represents a threat to what might be thought of as a spatial contract between neighbourhoods of differing social characteristics. As private governance has grown, the concentration effects of poverty and problems of crime displacement have been systematically layered onto an urban poor already weighed down by a wide range of local problems. In other words, GCs can be seen as undermining local and central state responsibilities to at least attempt to create equity of outcome between neighbourhoods of different social characteristics and qualities. This is such a central concern of much writing on GCs that it is surprising that so little hard evidence exists.

Most commentators consider the effect of GCs to be socially divisive, while it has been suggested by some in the British policy community that the security features of GCs could tempt owner occupiers to 'colonise' more deprived neighbourhoods, thus creating mixed communities or what might be better termed 'citadel gentrification'. However, results from an analysis of the 2001 American Housing Survey, by Sanchez *et al.* (forthcoming), are somewhat surprising. They found a prevalence of low-income, racial minority, renters living in GCs. The extent to which US public housing projects are walled and gated is illustrated by results showing that tenants are nearly 2.5 times more likely than owners to live in GCs. The study provides a baseline for tracking the racial and income mix within American GCs over time. This challenge to the image of private affluence, underpinned by an apparent democratisation of gating, is perhaps no less depressing in its impression of cellular and parallel lives lived in, and driven by, fear.

Joining the Club: The Drivers and Governance of Gated Communities

The contributions to this volume come from across the globe, with papers from Brazil, Taiwan, Argentina, Germany, China, Canada, the UK and the US, highlighting both the widespread proliferation of and interest in GCs. The authors offer empirical data on the effects of GCs in different locations and political contexts, and many of the contributions address more than one of the themes outlined in this introduction.

One major concern is to provide a convincing account for the growth of GCs. Webster has previously drawn on economic theory to explain the way in which residents group together to provide services which the state can no longer adequately deliver, suggesting that private neighbourhoods have much in common with public government at the local level (see for example, Webster, 2002). A closer reading, in this volume, of the history of Taiwanese homeowner associations by Chen and Webster leads them to conclude that these organisations suffer from many of the same problems as conventional municipal government. The contribution from Wu here finds that club goods theory, rather than flight from fear, provides an explanation for the transition from work-unit compounds to gated commodity housing enclaves in urban China, another example of what Wu here terms 'the worldwide restructuring towards entrepreneurial governance'.

Glasze here develops these themes, arguing that club goods theory is insufficient to explain the historically and regionally differentiated development of GCs. He suggests that the crucial question of power has been overlooked, and further that club goods theory is inadequate when evaluating the potentially divisive consequences of GCs, which naturally tend to look after the interests of their members only. Glasze further proposes that more attention should be paid to the role of the state, on a national, regional and local level, when attempts are made to explain the rise of gated developments. Evan McKenzie (1998) has previously suggested that this is dependent on the relationship between three vectors: the developer, potential residents, and 'the municipality', with multiple forces acting on each of the three, so that outcomes are affected, for example, by the price of land or the relative affluence of the municipality. In his paper, McKenzie applies his analysis to the rapid development of GCs in and around Las Vegas, where local government takes the lead in promoting GCs because they represent growth, increased tax revenues and less public expenditure. The municipalities now require new developments to be managed by homeowner associations and, while property developers are happy to go along with this, purchasers have no choice in the matter.

Discussion of the relationship between these three vectors again emerges in other papers in this volume; for example, Chinese local authorities accept GCs as they relieve the municipality of its administrative and public service burden (Wu), whereas Thuillier shows how Argentinian GC developers and residents are able to play off competing municipalities against each other, and keep down local taxes. We find in Chen and Webster's paper that the Taiwanese government now has the power to insist on homeowner associations as the accepted form of governance in new developments, as in McKenzie's analysis of Las Vegas. There are no examples here of successful government control of GCs. Again Grant's work here shows Canadian planning authorities to be struggling with the same issues as English planners (see Atkinson *et al.*, 2004, and Manzi & Smith-Bowers, this volume), in a context where GCs are a new, and rare, phenomenon not (yet) aggressively marketed by property developers. For very different reasons, local government in Argentina appears powerless to control the dynamic growth of GCs (Thuillier).

Contractual Neighbourhood Governance and Citizenship

We have argued that the contractual legal framework is an essential characteristic of GCs. These detailed rules indicate a different and much more formal structure than the framework of informal rights and rules developed in a neighbourhood through "neighbours understanding the importance of maintaining a shared and reciprocated set of values and neighbourhood attributes" (Webster, 2003, p. 2606). It has been suggested that GCs are an example of a much wider rise in contractual governance, resulting from the new relationship between state, market and civil society, designed to address concerns about social order: the contract of membership takes centre stage in the age of 'responsibilisation', in which "exclusion from club goods may be tantamount to exclusion from key aspects of citizenship" (Crawford, 2003, p. 500).

The contractual nature of GCs implies a sense of choice by each party entering into a contract, voluntarily accepting its future restrictions, whereas research shows that GC residents appear unaware of the details of what they are signing up to. A 1989 survey in 12 Californian counties found that only 27 per cent of resale purchasers had read the covenants and then only when accused of breaching them (Silverman *et al.*, 1989). In another American study, less than 10 per cent of residents had read the legal documents prior to purchase (Alexander, 1994). Similar findings appear in two of the contributions in this volume. In their study of an English GC, Blandy & Lister found a general lack of awareness of the details of the legal framework for self-management by residents, and ignorance of (as well as indifference to) the covenants. Here, McKenzie describes the rules of American homeowner associations as "incomprehensible and non-negotiable". He concludes that, far from a voluntary community bound by contract, GC purchasers have an enforced relationship with their homeowner association. It is only if the conceptual framework of contract is stretched to breaking point that GCs, at least in England and the US, can be seen as a form of genuine self-management. Crawford tacitly acknowledges this when he includes in his examples of contractual governance those which "have emerged in part as problem-solving devices" (Crawford, 2003, p. 481).

In the case of GCs, it seems that the provisions of property and contract law have been used to facilitate a built form that appeals to both developers and residents. However, one party frequently enters this particular contract either because there is no alternative, or at least in ignorance of the legal implications. This provides a challenge to club goods theory in which, far from being the rational choice of economic actors maximising security, buying into the 'community' may be a simple lifestyle preference in which the trappings of onerous legal details and coercive clauses are simply ignored. While the economics of the context are understood by the various players, it is in the social aspect of the model that residents find themselves internally 'outcast' for infractions like flying a flag on the wrong day or having the wrong size or type of car. Many potential residents may fail to anticipate that the price of 'total' security is the loss of many such minor liberties.

In his paper, Glasze characterises GC homeowner associations as 'territorial shareholder democracies' and points to a number of difficulties that may arise in this privatised model of governance. Two papers in this volume also consider whether social capital within the development is enhanced by the fact that GC residents are contractually bound together. Blandy & Lister's paper suggests that residents see covenants as a substitute for community controls and sanctions, and that the 'community' of residents is characterised by weak ties. In his analysis of GCs in Las Vegas, McKenzie argues that the legal governance arrangements do not encourage a localised identity, but that the GC operates more as an economic collective which provides services traditionally seen as the task of local government. Chen & Webster address similar issues in their application of Olson's theory of the problems of collective action to homeowner associations in Taiwan. Although resident participation in Taiwanese GCs is specifically promoted by property management companies, there is evidence of a general reluctance to join resident management boards. In the new Chinese private gated enclaves, Wu's contribution shows that residents' desire for anonymity, and reluctance to participate in 'unnecessary' social interaction, means that management and governance are largely left to professional property companies.

As a result of these common features of the legal framework, and residents' attitudes to them, GCs worldwide are prone to internal dissent and vulnerable to the possibility that a powerful minority may impose its will on the majority (see McKenzie, this volume). In recognition of these problems, a number of countries are now introducing regulation of homeowner associations, for example, requiring them to hold minimum sinking funds or to hold regular elections. McKenzie's paper sets out the different approaches taken by three state legislatures in the USA. This raises the possibility that an interesting cycle may be emerging whereby privately governed neighbourhoods develop to provide goods which local government is inefficient at providing, but the private regime then becomes oppressive to the residents who are 'club members', and itself has to be curbed by government legislation.

The theme of segregation, and the effect of physical boundaries in the differentiation of 'them and us', is addressed by Roitman and others here. Both Thuillier's and Roitman's papers examine fortified GCs with fearful affluent occupiers, in the context of a highly polarised society in Argentina. Moving to the west coast of the US, Le Goix presents the first large-scale analysis of segregation patterns in GCs and their surrounding neighbourhoods by integrating a GIS database with data from the 2000 Census on more than 200 GCs in the Los Angeles area. He finds that GCs have a negative impact, causing what he terms the reinforcement of 'segregation within segregation', especially relating to age and socio-economic status.

Finally, Manzi & Smith-Bowers provide empirical evidence from two case studies in London, arguing that gates and other security features are needed to convince owner occupiers to move into (and remain in) areas where the fear of crime is high. While their presence appears to reduce segregation at the neighbourhood level, the reality is a divided area; interviews carried out by the authors indicate that the GC residents do not mix at all with residents outside the gates. This contribution stands out in its relatively positive reading of the potential for gated communities to make some contribution to social integration, thereby providing a counter-point to the generally negative impressions given by other authors.

Conclusion: Liberated Enclosure?

Gated communities represent a new or at least relatively novel form of housing development in the European context and their number is increasing. With growing consumer and media interest the US and South African models of such development may form templates for understanding this direction in preferences, primarily directed by fear, privacy and predictability. What is less clear is why such development is growing in societies characterised by lower prevailing crime rates and higher levels of social cohesion. In this sense perhaps gated communities might be seen as barometers indicating the future shape and scale of social forces linked to social fear and aspirations toward ex-territoriality (Bauman, 2000). In this sense the significance of gated communities lies less in their number and more in what they say about a wider bundle of social pressures now directing where and how people live.

For housing researchers, gated communities intersect important threads of urban theory and empirical profiling as well as normative and political ideas about what kinds of housing we want to see provided in our towns and cities. Certainly for many academics gated communities have provided a rich vein for research since they conflict with the personal politics and wider ideals often enshrined in planning frameworks as well as attempts at achieving relative social justice and balance in the neighbourhood context. Gated communities are the clearest indication that unimpeded consumer and developer choice threatens these wider aspirations. The appeal to legitimacy by supporters of gated development has been based on the notion that people's preferences are increasingly based on fear because the state has not fulfilled its contract in delivering safety. However, as we have argued here, these choices are not without wider impacts and have been shown to amplify personal fears, crystallising patterns of segregation, and displacing crime.

The withdrawal of the generally affluent into gated enclaves presents us with a range of possibilities. First, a loss of social diversity in the neighbourhoods that they have left leading to a residualisation of exited locales, thus reinforcing tendencies toward social segregation. Second, the displacement of crime away from increasingly hardened targets, inhabited by those who can afford access to security, towards those areas which present softer targets. Finally, gating represents a spatial withdrawal of elite groups that threatens what we have described as a spatial contract between neighbourhoods in cities mediated by central and local states. Services for poorer areas may suffer as a result of the opting out of municipal provision that has been argued for by many gated communities in the US, which have their own privatised fiscal arrangements and revenues.

This brings us back full circle to arguments about the purpose of central civic bodies which collect and disburse resources to the benefit of all members of a society. The club good of security and neighbourhood services represented by gated communities resemble new medieval city-states wherein residents pay dues and are protected, literally as their 'citizens'. With the growth of these gated mini-states, the argument has been that gated residents should not have to pay twice for services they already receive. This may ultimately have the effect that entitlements to vital aspects of citizenship, such as security, welfare and environmental services, become based on which neighbourhood one lives in. With concerns currently running high about 'postcode (zipcode) lotteries' relating to educational, welfare and health service access, these problems look set to take on a greater analytical significance in the future.

The papers in this volume reflect on these and other issues and provide a range of both explanatory and descriptive frameworks for trying to understand why gated communities have arisen, why they persist and whether or not normative theories of the good city should challenge gated neighbourhoods as desirable in a context of wider social justice in the city. No doubt this volume will not be the last word on what continues to present a controversial topic for analysis by urban and housing researchers.

References

Alexander, G. S. (1994) Conditions of 'voice': passivity, disappointment and democracy in homeowner associations, in: S. E. Barton & C. J. Silverman (Eds) *Common Interest Communities: Private Governments and the Public Interest* (Berkeley, CA: Institute of Governmental Studies Press, University of California).

Atkinson, R., Blandy, S., Flint, J. & Lister, D. (2004) *Gated Communities in England* (London: Office of the Deputy Prime Minister).

Atkinson, R. & Flint, J. (2004) Fortress UK? Gated communities, the spatial revolt of the elites and time-space trajectories of segregation, *Housing Studies*, 19 (6), pp. 875–892.

Atkinson, R. & Kintrea, K. (2002) A consideration of the implications of area effects for British Housing and regeneration policy, *European Journal of Housing Policy*, 2, pp. 1–20.

Barton, S. & Silverman, C. (1987) *Common Interest Homeowners' Associations Management Study* (Sacramento: California Department of Real Estate).

Bauman, Z. (2000) *Liquid Modernity* (Cambridge: Polity).

Berlin, I. (1958) *Two Concepts of Liberty* (Oxford: Clarendon Press).

Blakely, E. & Snyder, M. (1999) *Fortress America: Gated Communities in the United States* (Washington, DC and Cambridge, MA: Brookings Institution Press, Lincoln Institute of Land Policy).

Crawford, A. (2003) Contractual governance of deviant behaviour, *Journal of Law and Society*, 30, pp. 479–505.

Gans, H. (1968) *People and Plans: Essays on Urban Problems and Solutions* (New York: Basic Books).

Graham, S. & Marvin, S. (2001) *Splintering Urbanism: Networked Infrastructures, Technological Mobilities and the Urban Condition* (London: Routledge).

Hope, T. (2000) The clubbing of private security: the collective efficacy problem for rich and poor, in: T. Hope (Ed.) *Perspectives on Crime Reduction* (Aldershot: Ashgate).

Linford, J. B. (2001) *Planned Community Living: Handbook for California Homeowners Associations* (San Clemente, CA: Felde Publications & Programs).

Low, S. (2003) *Behind the Gates: Life, Security, and the Pursuit of Happiness in Fortress America* (London: Routledge).

McKenzie, E. (1994) *Privatopia: Homeowner Associations and the Rise of Residential Private Government* (New Haven and London: Yale University Press).

McKenzie, E. (1998) Homeowner associations and California politics: an exploratory analysis, *Urban Affairs Review*, 34, pp. 52–75.

Massey, D. & Denton, N. (1993) *American Apartheid: Segregation and the Making of the Underclass* (Cambridge, MA: Harvard University Press).

Minton, A. (2002) *Building Balanced Communities: The US and the UK Compared* (London: RICS).

Sanchez, T. W., Lang, R. E., & Dhavale, D. (2003) Security versus status? A first look at the census's gated community data, *Journal of Planning Education and Research*.

Silverman, C. J., Barton, S. E., Hillmer, J. & Ramos, P. (1989) *The Effects of California's Residential Real Estate Disclosure Requirements* (Sacramento: California Department of Real Estate).

Webster, C. (2001) Gated cities of tomorrow, *Town Planning Review*, 72, pp. 149–170.

Webster, C. (2002) Property rights and the public realm: gates, green belts, and Gemeinschaft, *Environment and Planning B*, 29, pp. 397–412.

Webster, C. (2003) The nature of the neighbourhood, *Urban Studies*, 40(13), pp. 2591–2612.

Wilson Doenges, G. (2000) An explanation of sense of community and fear of crime in gated communities, *Environment and Behaviour*, 32, pp. 597–611.

Constructing The *Pomerium* in Las Vegas: A Case Study of Emerging Trends in American Gated Communities

EVAN McKENZIE
Political Science Department, University of Illinois at Chicago, USA

(Received October 2003; revised March 2004)

KEY WORDS: Gated community, homeowners association, privatization, special district

Introduction

Privately governed residential enclaves, known as common interest housing developments (CIDs), many of them gated and walled, are the predominant form of new housing in America's fastest growing cities and suburbs. Over the last 25 years, this massive privatisation of local government functions has changed the appearance and organisational structure of American urban areas. This trend is not a passing fashion but an institutional transformation reflecting the ideological shift toward privatism characteristic of the neo-liberal consensus. Specifically, the CID revolution is driven by three main forces. Developers pursue higher density in order to maintain profits despite rising land costs. Local governments seek growth and increased tax revenues with minimal public expenditure. Many middle and upper-class home buyers, fearful of crime and disenchanted with government, are in search of a privatised utopia offering security, a homogeneous population, and managerial private government.

This transformation resembles the construction of a physical and institutional *pomerium*, or sanctified wall, around the affluent portions of an increasingly divided society. Nowhere in the US is this transformation more visible than in Las Vegas, Nevada, the fastest growing city in the nation, and one that exemplifies the national and global trend toward placing tourism at the centre of the urban economy and reshaping the spatial, social and political order accordingly. Las Vegas area local governments require developers to construct virtually all new housing in CIDs, and gated security developments are popular. So popular, in fact, that non-CID neighbourhoods come under pressure to emulate CIDs. One such neighbourhood, Bonanza Village, was literally walled in by the City of Las Vegas, over the protest of many of its residents, in order to make the old neighbourhood resemble contemporary gated communities and thus link it with downtown redevelopment.

This paper seeks to outline the trends that are emerging in the production and practices of these privately governed gated communities.

Production: Residential Private Government and Gated Communities

Privately governed residential enclaves, known as common interest housing developments (CIDs), are the predominant form of new housing in America's fastest growing cities and suburbs. About one-fifth of them are gated and walled (Blakely & Snyder, 1997, p. 180, n.1). Over the last 25 years this massive privatisation of local government functions, consisting of some 250 000 housing developments containing about 20 million housing units and 50 million people, has changed the appearance and organisational structure of American urban areas (Community Associations Institute, 2004).

Common interest housing includes planned developments of single-family homes, townhouses, and condominiums. These developments involve a form of ownership in which home buyers purchase both an individual interest in a particular unit and another interest, consisting often of streets, recreation centres, golf courses and other facilities, which they own in common with all residents in the development. They buy their property subject to voluminous sets of deed restrictions, rules and regulations, under which all owners agree to make monthly payments to a homeowner association, a private government into which all residents are enlisted at the moment of purchase. The association is run by the residents, supported by cadres of lawyers and other professionals, and it enforces the deed restrictions against all residents and manages the use of property and other aspects of life in the development. Increasingly, CID housing involves homeowner association-administered security measures, which typically include walls and gates, and may involve hiring guards and even private police forces.

There is considerable disagreement over the causes and effects of this phenomenon. It has been argued that the CID revolution is driven by the motivations of developers and local governments on the supply side, and consumers on the demand side, with the supply side interests predominating over the demand side (McKenzie, 1998a).

Developers have found that CIDs help them pursue higher density in order to maintain profits despite rising land costs. They can put more people on less land, and also provide amenities to buyers, by creating common ownership of parks, swimming pools and so forth. Local governments seek growth and increased tax revenues with minimal public expenditure. CIDs privatise what would otherwise be government responsibilities and place these burdens in the hands of homeowner associations, whose members pay for them through monthly assessments. These associations arrange for rubbish collection, plough snow in the winter and move leaves in the fall, repair and light streets, run parks and do many other things that government would otherwise have to do in order to enjoy the increased tax revenues from new development. Thus cities can acquire new property taxpayers without having to extend to them the full panoply of municipal services.

But the demand for such a lifestyle cannot be ignored. Many middle and upper-class home buyers, fearful of crime and disenchanted with government, are in search of a privatised utopia offering security, a homogeneous population, and small-scale managerial private government that enforces high standards of property maintenance. For many people, the gated community is especially attractive, as it adds fortification to all the other attributes of CID living.

I have argued that the rise of residential private government facilitates the emergence of a two-tier society in which the 'haves' are increasingly separated—spatially, institutionally, socially and economically—from those of lesser means. I call this realm 'privatopia' because it represents the pursuit of utopian aspirations through privatisation of public life. Within privatopia the terms and conditions of life are at odds with the norms and expectations of liberal democracy. Residential private governments, known generically as 'homeowner associations', are not restricted by conventional notions of civil liberties and due process of law, and their activities are supported by a powerful cadre of professionals, including lawyers, property managers, accountants and others.

Yet, many observers see the situation quite differently. Some argue that the CID revolution is merely a manifestation of consumer sovereignty, representing the collective preferences of millions of home buyers. This demand side logic reaches its greatest extent with the libertarian justification of homeowner associations as private protective associations, a view anticipated in Robert Nozick's major work, *Anarchy, State, and Utopia.* By this logic, discussion of the social effects of these millions of individual choices is largely irrelevant, because principles of individual liberty that govern the choices justify the end result. The related *caveat emptor* argument is generally persuasive to American courts, reflecting the view that each individual owner should be bound by the terms of her contract, and that the state should not interfere to remake that agreement. This argument has been considered elsewhere (McKenzie, 1998b), that the premises for the *caveat emptor* perspective often do not apply, because in many cases the contracts that create homeowner associations are in reality adhesion contracts, the terms of which are incomprehensible to the average buyer and non-negotiable in any event. However, it seems that the social and political consequences of private residential government are too significant to be left to individual market choices.

The homeowner association is not a passing fashion but an important institution, reflecting the ideological shift toward privatism that is characteristic of the neo-liberal consensus. Institutions insinuate themselves into people's lives, shaping the way they think and the choices they make. Mandatory membership homeowner associations induce people to identify with a small neighbourhood of people with similar social and economic characteristics, either by co-operating with the association or by opposing it. This is a kind of localised identity formation that otherwise might not happen. Some scholars, particularly those of communitarian leanings, like to think of this process as social capital formation, or as an embodiment of the 'defensible space' theory, and some think it is a voluntary community. The interpretation here is that typically this institution gathers a group of affluent people together and forces them to think of themselves in relationship to the institution and the neighbourhood it represents. It also locks them together economically to do things that otherwise local government would do. Although developers started this institution, in the last decade state and local governments have taken the lead in promoting the spread of CID housing.

What is the relationship of gated communities to this privatisation process and the institution that is at its core? Taken together, these things—homeowner associations, privatisation and gated communities—resemble the construction of a *pomerium*. The *pomerium* is an ancient concept dating to pre-Roman times and used in the demarcation of Rome itself. The *pomerium* was not necessarily a real wall, although it had physical markers. It was a symbolic, sanctified boundary that separated civilisation from barbarism, order from chaos and civil peace from anarchy. The *pomerium* was, in essence, an imaginary line drawn around the spiritual city. Instead of surrounding an entire city, today's emerging *pomerium* demarcates the protected islands of walled and gated private communities.

Practices: Homeowner Associations, Security Walls, and Development Trends in Las Vegas

Las Vegas is the fastest growing city, in the fastest growing county, in the fastest growing state in the USA. The spread of CID housing as the dominant form of new residential development is especially dramatic in the Las Vegas area. Nearly all new construction is in planned residential subdivisions with homeowner association private governments. In order to maintain low taxes with an astronomical growth rate, the City of Las Vegas and Clark County promote CID housing, which offers those who can afford it a range of privatised services, and minimises demands on local government. As Gottdeiner observes:

> While master-planned communities have been criticized as being insular for isolating themselves from the surrounding community, that is exactly what many homebuyers want ... In short, they seek services and protection they can no longer expect from municipal government. Thus, while some may criticize them as sterile, master-planned communities continue to be a great success in the Las Vegas region, where developers continue to build and sell thousands of homes per year. (Gottdeiner *et al.*, 1999, p. 153)

While there is clearly a demand for such locations, their proliferation is not just the byproduct of interaction between buyers and sellers. The City of Las Vegas virtually mandates that new development be done with homeowner associations. This is a two-step process. First, the city's Zoning Code and Development Code require that all new housing within the planned development zone contain certain features, including a landscaping plan, open spaces, and often security walls. Then, elsewhere in these codes, the city requires that *if* such features are included—which they must be—then there must be a homeowner association to maintain them. For example, in the following excerpt from Title 18 of the Las Vegas Zoning Code Section 18.12.5600, the word 'shall' was recently substituted for the word 'may' to provide as follows:

> *18.12.5600 Landscaping Plan. A landscaping plan shall be provided* by the subdivider as an integral part of subdivision design. Such a plan shall be prepared

> and submitted with each final map application addressing the landscape design of the subdivision with respect to such features as wall or fence design; land forms or berms; rocks and boulders; trees and plant materials; sculpture, art, paving materials, street furniture; and subdivision entrance statement; common area landscaping and other open space areas ... *Where common lots are shown for landscaping, the applicant shall cause the creation of a homeowners association for purposes of owning the common lot and maintaining the landscaping.*

The code further provides that "All walls, setback areas and landscaping created to accommodate these regulations shall be located on private property. *If in common ownership, the property shall be owned and maintained by a Homeowner's Association*" (Las Vegas Zoning Code, Section 18.12.570, subsection C). And Chapter 19 of the Zoning Code requires that in Residential Planned Development Districts, "All development with 12 or more dwelling units shall provide 15 per cent useable open space for passive and active recreational uses".

The city's Urban Design Guidelines and Standards are similar, stating: "All required landscaping shall be properly maintained, based on standard landscaping practices, by the property owner(s) and/or *supported by a perpetual Homeowner's Association budget*, or a reasonable alternative approved by the City". According to a representative of the Southern Nevada Builders' Association, no such alternative has been approved to date. The same Guidelines and Standards provide that "Developers may provide and plant street medians on public and private streets as long as they are supported by a perpetual Homeowner's Association". Elsewhere, common open spaces, which must be HOA controlled are required: "Private and *common open spaces are to be provided* in Residential Planned Development Districts and in multi-family residential developments".

Title 19 of the city Zoning Code provides for HOA controlled private streets and gated entrances:

> Subdivisions developed with private streets must have a mandatory property owners' association which includes all property served by private streets. The association shall own and be responsible for the maintenance of private streets and appurtenances ... The entrances to all private streets must be marked with a sign stating that it is a private street. Guard houses, access control gates and cross arms may be constructed. (Chapter 19A.04)

Las Vegas Mayor Oscar Goodman was forced to respond to these "complaints that ordinances on the books since 1997 mandate all new subdivisions be structured as homeowner associations", saying to angry builders and owners only that, "We can see if we can make some adjustments" (*CityLife*, 2000). But such adjustments are neither forthcoming nor probable. The City of Las Vegas not only requires HOAs in new development, but also encourages existing neighbourhoods that do not have homeowner associations to form them. In 1998, the Las Vegas City Council unanimously approved a measure directing city staff to work with neighbourhood 'community associations' in "crafting plans to guide development" in the city (Zapler, 1998) Through the Neighbourhood Services Department, the city has induced over 150 different neighbourhood associations to form (Gottdeiner *et al.*, 1999, p. 182).

There is one other major ingredient driving the current political economy of Las Vegas, and that is the competition between the City of Las Vegas and Clark County for tourist dollars. Downtown Las Vegas, known as 'Glitter Gulch', was the home of the original Las Vegas casinos. But over the last 20 years, these casinos have been eclipsed by the construction of giant, spectacular 'mega-casinos' on Las Vegas Boulevard, or 'The Strip', outside the city limits. These mega-casinos are close to downtown and closer to the major airport, but are in Clark County. They drain tourist revenue from the city, creating a uniquely intense version of the city-suburb competition for business that is typical of most American metro areas. During the 1990s, the City of Las Vegas fought back against the mega-casinos with a massive downtown redevelopment effort. Over half a billion dollars in development funds are being channelled through the Center City Development Corporation (CCDC), a non-profit corporation that is a 'private-public partnership' modelled after the entity used to redevelop downtown San Diego, California.

As in other downtown redevelopment efforts, the poor stand in the path of the city's economic resurgence in its competition with the county. The city recently forced one of the area's largest homeless shelters out of Las Vegas, denying it title to 10 acres of land it had been operating on, just north of downtown. As the Mayor and a city council member said:

> "I don't want to see Las Vegas become the only center for the homeless in this valley," Las Vegas Mayor Oscar Goodman said. "This is a problem that must be shared with the entire region." Las Vegas Councilman Lawrence Weekly said he doesn't want a "homeless Taj Mahal" built on the premises, where a shelter, medical clinic, crisis intervention facility and job counselling center have operated since the San Diego-based charity expanded to Las Vegas. (Moller, 2001)

Councilman Weekly's ward includes not only the homeless 'Taj Mahal' he spoke of so compassionately, but also Bonanza Village. As will be seen, the Bonanza Village episode illustrates the dynamics just described.

Case Study: The Bonanza Village Wall

Bonanza Village is a development of single-family homes in what is known as the 'West Side' of the City of Las Vegas. The development was established in 1946 with 168 lots. At that time the area was on the western edge of the city, bordering on the desert. However, over the years the city has spread so far to the west that now the so-called West Side is actually located near the centre of the city. The West Side is the historically black area of Las Vegas, although in recent years there has been an influx of Hispanic residents. Bonanza Village is located about 1.5 miles northwest of downtown Las Vegas, or 'Glitter Gulch', the location of the original casinos that is currently undergoing large-scale redevelopment.

Bonanza Village is bordered by four streets: on the north by Owens, on the east by Martin Luther King, Jr., on the south by Washington and on the west by Tonopah. As originally constructed the development could be entered by automobiles at five places from these public streets, from all four compass points. It was, and is, zoned 'R-E', or 'Residential Estates'. This classification, sometimes called 'horse lots', bears the official description of providing for "low density residential units located on large lots and conveying a rural environment". The lots are over half an acre in size. There were restrictive covenants, including a whites-only race restrictive covenant, recorded against the lots at the time of original construction, but, as was normal for that era, these covenants did not create a homeowner association to enforce them. There are no private streets or other 'common area', meaning property owned jointly by all residents. The covenants allow residents a great deal of latitude in the use of their land as long as it remains a single-family dwelling. They are allowed to have various kinds of outbuildings including stables, guest or servants quarters and greenhouses; grow crops for wholesale as long as they do not put up advertising signs; and have a 'reasonable number' of animals, including dogs, cats and horses, and up to 12 chickens and 6 rabbits.

Many of the houses in Bonanza Village were built in the 1950s and 1960s, and a substantial number are still occupied by the original owners. These are older people, many living on fixed incomes. However, over the last two decades, there has been an influx of younger, more affluent residents to Bonanza Village. At present, the development is more or less evenly divided between these two groups (Edwards, 2000). The newer residents envisage remaking Bonanza Village in the mould of the newer subdivisions, with a homeowner association, walls and a gate, and a high standard of property maintenance. Given the large lot sizes and proximity to the redeveloping downtown, they anticipate that a substantial increase in property values would result if they could make the property fit the expectations of home buyers like themselves—young, professional and racially diverse. But older residents are accustomed to the freedom and the relaxed, near-rural lifestyle that Bonanza Village has always had, and they are fearful that they will lose their homes through the increase in property taxes and maintenance costs that could result from gentrification of their development.

The dispute over the Bonanza Village wall is in large part a conflict between these two groups, and the wall came to symbolise the gentrified Bonanza Village envisioned by the newer residents and feared by the older ones. The dispute began in the early 1980s, and at every turn the wall advocates identified themselves as a voluntary homeowner association.[1]

Efforts to close off Bonanza Village from the surrounding areas were in evidence as early as 1981, a group of residents calling themselves the Bonanza Village Homeowners Association successfully petitioned the City of Las Vegas to vacate four of the five intersections from which Bonanza Village could be entered, leaving only the entrance from Washington at Comstock on the south of the development.

The city conditioned its surrender of these streets on the homeowner association installing 'crash gates' at three of the vacated intersections and keeping them locked at all times but available for emergency access. These gates were never installed, and instead the association put concrete traffic barriers in place. The net result of this episode was that the city allowed the notional homeowner association to close the development to vehicle traffic except from the south. Pedestrians could still walk into the development at these intersections. Some residents protested the action, but to no avail (Hawley, 1982; Null, 1981; Ogilvie, 1982).

Much of Bonanza Village's perimeter had been enclosed over the years by walls or fences erected by most of the 65 lot owners who lived on the perimeter of the development. With these barriers to entry, and the closed streets, there was enough security and separation from the surrounding neighbourhood and security for many residents. However, others were not satisfied with these measures and they began to campaign for construction of a full perimeter wall.

In 1985, the Bonanza Village Homeowner Association—an unincorporated and voluntary organisation—requested creation of a Special Improvement District in order to build a wall around the development. A letter from the association to the City Attorney in 1985 requested creation of a SID for "construction of a masonry wall on the perimeter of Bonanza Village at locations where said wall is non-existent". The wall and SID, said the association, would produce "improved security of total Village area", "improved property value", "improved safety for all residents", "environmental improvement (reduction of noise level; reduction of vehicle traffic"), and "better control of unwanted foot traffic" (Simon *et al.*, 1985). This effort failed.

In 1990, a group of the newer residents attempted to secure passage of an amended set of deed restrictions that would have created a mandatory membership homeowner association for Bonanza Village. An incorporated association with power to enforce restrictive covenants would have been able to finance a wall with homeowner assessments and build a wall without petitioning the city to create a SID. However, the effort encountered sufficient opposition from other residents that it failed. Thereafter, the pro-gentrification residents returned to the earlier approach of using the voluntary homeowner association to petition the city for creation of a special improvement district to finance the wall construction (Wills, 2000).

In 1997, the effort began to gain momentum. A new president of the Bonanza Village Homeowners Association—which was still not incorporated—introduced herself to the membership in January of that year and announced what would prove to be the pivotal event that would make the association's wall a reality: the City of Las Vegas now wanted the wall to be constructed:

> I look forward to working together to make a reality that much talked about block wall. *The City has re-committed itself to the block wall by assigning a full time staff person to help us achieve our goal*. The individual will be working with us from the Neighbourhood Preservation Office on Owens. This is a new and exciting development and shows a commitment we've never had before. *There is much development going on around us* as well as inside. (There are two new homes currently under construction in the Village.) As I see it, right now is the best time for all of us to pull together to get our wall completed. I truly hope that you will make a commitment to become more involved in improving Bonanza Village in 1997. Yes, it's been difficult, yes, we've gone down this road before, but I say that's no reason to stop trying! *We as homeowners are responsible for what Bonanza Village looks like and what it is. If we really want to look better, IT'S UP TO US to clean it up*. I'm willing to take responsibility for that and all I need is YOUR HELP. If the City doesn't come through for us, I say we come up with another plan and do it ourselves! Let's do more than talk, *let's work together to save Bonanza Village*. WE CAN DO IT!. (italics added; capitalisation in original) (Bonanza Village Homeowners Association, 1997a)

The City of Las Vegas was, indeed, now supporting the wall project, with the city councilman for Bonanza Village's ward now pushing the project through. This time the project made sense to a city that was determined to redevelop and gentrify its downtown areas to keep pace with the County-based Strip. Bonanza Village lay within the West Side areas added in 1988 and 1996 to a special district called the Downtown Redevelopment

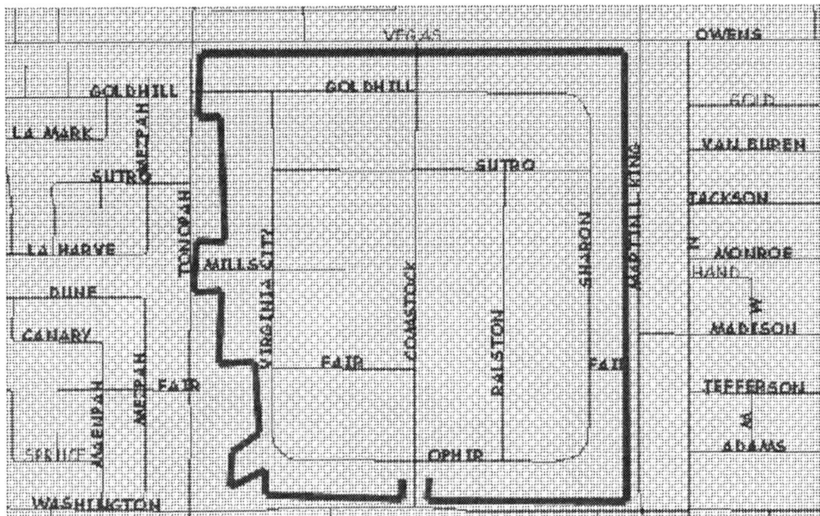

Figure 1. Bonanza Village neighbourhood showing approximate location of wall.

Area. This district was receiving special treatment from the city through the Center City Development Corporation, a non-profit corporation that was co-ordinating half a billion dollars in downtown redevelopment efforts. The Bonanza Village wall, and other efforts to make the old neighbourhood look like a contemporary gated community, were consistent with the city's overall redevelopment campaign. The wall opponents were no longer just opposing a group of their own neighbours. They were now, literally, fighting city hall. The association notified the residents in April 1997 that:

> The city now appears to be eager to get this project completed. Councilman Reese is applying pressure to the department of public works and other city offices to make them 'get on the stick' and quit procrastinating ... A 'special agent' has been appointed whose job it is to see that difficulties encountered in the planning and completion of the wall are quickly resolved ... The department of public works has completed detailed drawings of the perimeter wall. The wall being planned will be 8 feet high. (Bonanza Village Homeowners Association, 1997a, b)

This communiqué also revealed that the association's plans for Bonanza Village's makeover with the gated community image went beyond a mere 8ft wall: "We are planning to do something nice to the Comstock entrance as a part of the wall project ... We would like to get some kind of nice sign out front saying Bonanza Village and *there has been talk of a guard house being built in the island.*"[2]

There were two kinds of homeowner approval needed. First was the requirement of Nevada Revised Statutes Chapter 271 that SID creation be supported by a petition signed by 66.67 per cent of the owners who would be assessed. During 1997, the association went about the project gathering signatures on a petition to create the SID. The petition was drafted by the city for this purpose. There was major controversy over whether this requirement was satisfied, and anti-wall forces claimed the signatures had been collected over many years, were obtained with false information on the cost of the wall, included signatories who no longer lived in the development, and included others who had changed their minds. By September 2000, 86 complaints from anti-wall residents—a majority of the current residents of the development—were sent to the state Ombudsman for Common Interest Communities. The complainants asserted that "their subdivision was illegally organised . . . without the homeowner's knowledge", and that the protesters "do not want to be represented by the Bonanza Village Homeowners Association" (Ashworth, 2000). But the city and ultimately the courts decided that the petition requirement had been satisfied, regardless of these protests.

However, there was another set of signatures needed. The chief impediment to what was now a joint effort of the city and the homeowners association proved to be the need to obtain easements over the 65 properties that lay on the perimeter of Bonanza Village, where the wall would be constructed. The city needed a permanent easement for the wall itself, and also a larger temporary construction easement, and these required intrusion on private property rights. The homeowner association worked to obtain signatures on both the SID petition and the easement documents. But ultimately 13 of these perimeter owners did not sign the easement documents in time, and this precipitated the litigation that eventually took the Bonanza Village case to the Nevada Supreme Court.

The city set up an aggressive schedule for creating the SID and building the wall, all of which was to be done between February 1998 and February 1999. One of the items on the schedule was 'Protest Disposal Resolution', set for 11 May 1998. After the rejection by the city council resolution of the anticipated protests, the ordinance creating the SID was to receive its first public reading (*Bonanza Village Times*, 1998).

The schedule had to be set back a while and the protests were not rejected until 12 October 1998. The president and vice president of the Bonanza Village Homeowner Association addressed the Las Vegas City Council at a public hearing on 28 September 1998, to advocate for the wall project. They began by arguing that the wall was a crime prevention measure needed to prevent petty burglaries committed with shopping carts. The facts were that the crime rate in Las Vegas was plummeting, paralleling a similar drop in crime rates across the nation,[3] but crime was given as a justification nonetheless. They then portrayed the wall as an esthetic measure that would make Bonanza Village look like other, newer, communities, and thereby make the neighbourhood look like part of the downtown redevelopment. This, they said, would stimulate community pride and enhance property values. The president of the association spoke first (Las Vegas City Council, 1998):

> *Our reasons for wanting a security wall around Bonanza Village are the same reasons security walls have been and are currently being built, in fact, it seems to be the norm nowadays,* in many Southern Nevada neighbourhoods surrounding ours, like Summerlin, Green Valley, The Lakes, and Rancho Circle. *We want to discourage crime; namely, petty burglaries that occur on foot with the aid of a shopping cart. It's hard to push a shopping cart filled with a television or a lawn mower through a block wall, but this is the kind of crime that is occurring in our neighbourhood. So, like our surrounding neighbours, we wish to deter these crimes as well as protect our investments. We want to maintain a sense of neighbourhood pride and improve the appearance of the whole area.* Just like those other homeowners, we expect certain improvements for our tax dollars. Further because

> we are one of the oldest neighbourhoods in Southern Nevada, meaning we have been paying taxes for a longer time, we need to know that the City has not turned its back on us because of the new kids on the block. *Bonanza Village lies smack dead in the middle of a major downtown redevelopment that we want to be a part of.* There is the Fourth Street Corridor project under way, the old Union pacific Railroad land that now has a beautiful new County building in place, and right on Bonanza the old Dula City Swimming Pool is being brought back to life. *Why should our neighbourhood be left behind? The approval of the SID will breathe new life into the Martin Luther King/Washington/Vegas Drive area with the building of a security wall and will serve as an example for others in our area to show what neighbourhood pride can do and make—can be done to revitalise an old and impoverished area. It will stimulate neighbourhood pride and make Bonanza Village a part of the new downtown.*

The president was followed by the association vice president, who added to her comments, noting among other things that the city, not the association, had prepared the petition that the HOA circulated to obtain the signatures needed to proceed with the SID, and then proceeding to advocate for the wall purely because it would make Bonanza Village look like nearby gated communities:

> There is some new development that's happening in our area that we're excited about. The Andre Agassi Boys and Girls Club looks really good. The Veterans Hospital close by us looks good. There's a new Post Office going in. And also *there's a subdivision that has planted itself north of us. It's single family homes and they have a really nice wall running along Vegas Drive. Now all these good looking projects, when they look across the road or into our area, they see what typical horse zoned property, the back side, lots of typical horse zoned property looks like. It's a mish-mash of different fencing.* Sometimes you're not going to put your lumber pile in the front yard if you got a half acre. That's one of the reasons you own it is so that you can have a few things other than just landscaping. And so a lot of that stuff winds up in the back yards . . . The project is not only good for Bonanza Village, but it's good for the City. *If we do not build this wall, Bonanza Village will fall behind the level of the new developments coming into our area and we don't want that. We're an inner-city neighbourhood that's willing to spend our own money to bring itself up to the level of the new stuff that's moving in around us.*

A third wall supporter made the case even more bluntly, saying, "Today we would like to see Bonanza Village in the same capacity as any other gated community".

Those who objected to the wall spoke at this hearing as well, but five weeks later, on 9 November 1998, Special Improvement District 1463 was created. The degree to which the city was willing to overlook obstacles in its rush to build the wall was illustrated by an indemnity agreement which was entered into on that date between the City of Las Vegas and the Bonanza Village Homeowners Association, 'a Nevada corporation'. In fact, the association was not incorporated, and did not incorporate until nearly a year and a half later, on 7 April 2000, with only 22 members.

The fact that the city was entering into a contract with an organisation that had no corporate existence did not stop the city from moving forward. The indemnity agreement set out the arrangement for financing, building and maintaining the wall.

Bonanza Village became Special Improvement District No. 1463 for purposes of building the wall. All property owners within the development were to pay equal shares for construction of the wall, pursuant to state statute. Other documents show that the brown, 8ft high, cinder block wall—two miles of it in all—was expected to cost over $800 000, and that each lot would be assessed approximately $5000 to pay for it (the final cost to be determined after the wall was built), with up to 10 years to pay the principal and interest.

After the wall was constructed, the association was to be fully responsible for maintaining the wall, and they also agreed to defend and indemnify the city against any claims made arising from the existence of the wall. The president of the association signed the indemnification agreement for 'Bonanza Village Homeowners Association, Inc.'—the corporation which did not exist—as did the Mayor Pro-Tem on behalf of the City of Las Vegas. The indemnification agreement was then recorded against every lot in Bonanza Village, as though it were a mandatory membership organisation that spoke for all residents, when in fact it was voluntary and even as late as 2000 only had 22 members. This raised the possibility in the minds of some owners that they would become individually responsible for liability under the indemnification agreement, given the fact that the voluntary homeowner association had no means to raise money except by asking for it from residents and hoping for the best.

Later, on 4 July 2000, a Bonanza Village homeowner wrote to city Councilman Larry Weekly, the new representative for their ward, to point out the problem of the non-existent association, and asked, "Doesn't the city require documentation that a Homeowners Association is a Bona-Fide one before awarding $825 000 for a SID?" After some six weeks had passed, one of Weekly's employees passed the letter on for reply to the Supervisor of the Special Improvement District, who responded by saying that, "The fact that a homeowner association exists in a neighbourhood has nothing to do with the SID process. Their only involvement in this case was to walk the petition around to get the supporting signatures to start the process" (Thompson, 2000). Yet, the association's involvement in the wall project had predated the SID process, and would post-date it as well given the indemnification agreement.

Armed with the Special District legislation and the indemnification agreement with the association, the city proceeded against the Bonanza Village perimeter residents who had not granted the city easements over their property. Foremost among these was Cuthbert Mack, an attorney. Mack and other opponents of the wall continued to organise and fight against the wall. When it became clear that a number of perimeter property owners would not voluntarily sign away the easements, the city decided to exercise eminent domain and take their property.[4] On 28 June 1999, the City Council authorised filing complaints in eminent domain against the recalcitrant owners, and on 19 October 1999, the case of *City of Las Vegas v. Mack*, Case No. A410116, was filed in the Clark County District Court.

Mack's argument was that the city lacked authority to take his property for the wall project, and the eminent domain action violated the Due Process Clauses of the US and Nevada Constitutions. His central contention was that there was in reality no public use involved, and that his property was being taken by the city for what amounted to a private use. He also argued that there was no objective need for the wall, there was no need to take his property to construct a wall, and that the city had failed to follow the statutory requirements for SID creation, including lack of notice (*City of Las Vegas v. Mack*, Case No. A410116, Motion to Dismiss Plaintiff's Complaint, p. 2, lines 13–19). In a separate suit (*Mack v. Bonanza Village Homeowners Association*, Case No. A421503), Mack also requested an injunction to stop construction of the wall until the case could be fully adjudicated, as well as declaratory relief.

On 31 January 2000, the District Court denied Mack's motion to dismiss the eminent domain complaint. On 8 April Mack petitioned the Nevada Supreme Court for a writ of prohibition. On 3 May the city awarded the wall construction contract to a builder, but on 16 June the Nevada Supreme Court issued an order staying the project until it could consider the merits of the case.

It was at this point that the city revealed just how determined it was to build the wall. The city took the position that the Supreme Court's order only applied to the Mack property, and not the rest of the development. The city ordered the contractor to proceed with the construction, which they did. Existing fences and walls were bulldozed, trenches were dug, temporary chain-link fencing was erected, and piles of cinder blocks and rebar were stacked all around Bonanza Village. Fire hydrants, water main valves and telephone poles that serviced Bonanza Village were walled out of the development.[5]

On 25 July Mack requested a contempt citation from the Supreme Court to punish the city for violating the order. On 3 August the Supreme Court issued another order reaffirming the earlier one, and making it clear that the order to stop construction applied to the entire project. At this point, the city halted the project.

The Supreme Court made its third and final ruling in the case on 15 March 2001, denying Mack's request for a writ of prohibition. The court ruled that Mack had failed to file a timely written objection to the SID as required by NRS 271.305, and therefore had

Figure 2. The wall during construction, at Tonopah and Goldhill, from outside Bonanza Village looking in. *Source:* photo by Monica Caruso.

waived his right to object and did not have standing to challenge the formation of the district. On the merits of his claim, the court held that:

> The district court did not exceed its jurisdiction by denying petitioners' motion to dismiss the City of Las Vegas' eminent domain complaint because the right of eminent domain is an attribute of sovereignty, the express provisions of NRS Chapter 37 and 271 authorize the city to condemn property for local improvements, and *a security wall is defined as a local improvement* pursuant to NRS 271.203. Thus, the city's formation of a special improvement district (SID) and its efforts to condemn petitioners' property for purposes of constructing a security wall are not unconstitutional. (Order Denying Petition for Writ of Prohibition, *Mack v. Eighth Judicial District*, Case No. 36091; italic added)

The language in italics in the quotation above reflects a provision of Nevada law which specifically defines a security wall as a 'local improvement', so that the city was not required to produce any further justification—a security wall is, per se, a project with a public purpose. The statute defines how SID funded security walls are to be paid for, stating that walls benefit all residents equally, so payment is to be equal: "Because the protection afforded by a security wall benefits each tract in the subdivision ... the governing body may apportion the assessments for a security wall on the basis that all tracts in the subdivision share equally in the cost and maintenance of the project" (Nevada Revised Statutes Section 271.367).

With this Supreme Court decision, all opposition to the Bonanza Village wall was effectively crushed, and the city completed the wall. Cuthbert and Lois Mack were soon in danger of losing their home in foreclosure for refusing to pay their share of the cost for the wall. The Macks and a number of other residents did not pay, and their homes thus became subject to foreclosure by the city. In the Macks' case, the amount for which they were willing to lose their home over this issue was $1391, clearly a small sum in comparison with the value of the home (Couzens, 2003). The city sought a compromise to avoid foreclosure, but Mack and his wife continued to dispute the charge as a matter of principle (Silver, 2003).

This episode is replete with ironies and contradictions, one of which sits at the sole remaining entrance to Bonanza Village. The homeowner association obtained a grant from the city to put up a sign where the anticipated gate and guard house were to be built. It is a rock bearing the legend 'Bonanza Village, Est. 1946'. The sign invokes the long heritage of the community, symbolising the original Bonanza Village, where residents lived as they chose at the edge of the desert, with horses, chickens, sheds and crops. This was the very Bonanza Village that the new residents wanted to eradicate, and the homeowner association that erected the sign was the instrument for doing so.

The sign was paid for by the city that bulldozed the entire perimeter of the area, and this compounds the irony. The sign commodifies the area's semi-rural heritage as a selling point for a neighbourhood well on its way to becoming another gated, walled and homeowner association-controlled subdivision, against the will of many if not most of its residents.

The rhetoric of government responding to the wishes of its citizens is used to legitimise and obscure the fact that it is suing them to take their land. The contract with a corporation

Figure 3. Entry to Bonanza Village at Comstock and Washington, and rock sign. *Source:* photo by Monica Caruso.

that did not exist suggests that the incorporated homeowner association is so central to this type of activity that the city willed it into existence.

In *City of Walls*, Teresa Caldeira writes:

> All fortified enclaves share some basic characteristics. They are private property for collective use, and they emphasise the value of what is private and restricted at the same time that they devalue what is public and open in the city. They are physically demarcated and isolated by walls, fences, empty spaces, and design devices. They are turned inward, away from the street, whose public life they explicitly reject ... they belong not to their immediate surrounding but to largely invisible networks ... Finally, the enclaves tend to be socially homogeneous environments ... Fortified enclaves confer status. (Caldeira, 2000, p. 258)

What Caldeira found to be true in Sao Paolo could have been written of Bonanza Village. The affluent residents who are transforming their neighbourhood into a fortified enclave, and the city officials that did the work for them, are part of the invisible network of professionals who understand the significance of secure residential compounds. The drive to wall-in Bonanza Village was an attempt to enhance the status of the development and raise its property values. While lip service was paid to crime prevention, it also seems that the residents who formed the Bonanza Village Homeowners Association were seeking status. In this respect, they were merely emulating their counterparts in the developer-created homeowner associations all around them. And the city, speaking the same language, saw the value of imposing the wall on the community's recalcitrant residents.

Higher property values mean higher property tax revenues, and the wall, it was felt, would make Bonanza Village a neighbourhood whose appearance would contribute to the overall ambience of the redeveloping downtown area.

In this context, it may have seemed an easy and obvious choice to exercise eminent domain, take people's property from them by lawsuit and force, bulldoze their fences and yards, and even disregard a direct order from the state Supreme Court. The overriding consideration, it seems, was to extend the *pomerium* around this piece of real estate, to embrace these 168 lots within the arms of the sanctified, symbolic wall that separates civilisation from barbarism, civil peace from chaos, and us from them. Underlying the rhetorical superstructure of community betterment and neighbourhood empowerment, what happened at Bonanza Village exposes the new architecture of control that polices this boundary. This apparatus is comprised of special districts, homeowner associations, private-public partnerships and other special purpose, hybrid creations—a fusion of state, market and civil society in which the categories and concepts of liberal democracy are irrelevant and alien.

Notes

[1] In 1971, the Bonanza Civic and Homeowner Association was incorporated. This was not the organisation that engaged in the subsequent wall efforts described in this paper. It appears to have been organised around issues pertaining to the neighbourhood elementary school on the southwest corner of Bonanza Village. Its membership included areas outside Bonanza Village, all Bonanza Elementary School teachers were made honorary members, and the principal office of the corporation was the address of the elementary school. The incorporation papers describe the corporation's purposes as to "provide neighbourhood protection", "improve relationships between the Bonanza Elementary School and the Community", "promote neighbourhood beautification, maintenance, and improvement of property", "support youth activities and recreation within the community" and "improve public relations within the community boundaries and with other civic and political entities". This organisation's corporate status was permanently revoked on 29 August 1995.

[2] The association circulated an architectural drawing of the proposed entrance, complete with guard house, and also a place for school children to wait for the bus, because school buses in Clark County do not enter gated communities and children must go to the entrance of the community to be picked up.

[3] According to FBI crime statistics, the crime rate in Nevada dropped by 4.8 per cent between 1987 and 1997, when the wall project began to gain momentum. For Las Vegas, the Metropolitan Police report that between 1995 and 2000, there was an overall drop in property crimes of almost 38 per cent, including a 40 per cent reduction in burglary.

[4] In Mack's case, the city determined that the easement was worth $1225.

[5] On 8 February 2001, a project notice was put out for bid by the Las Vegas Valley Water District to conduce extensive work at Bonanza Village. The project includes, among other things, "installation of 31 new fire hydrant assemblies". The engineer's estimate for the project was over $1.6 million.

References

Ashworth, M. (2000) Letter from State Ombudsman for Common Interest Communities to Christine Monroe, Subject: Bonanza Village, 15 September.

Blakely, E. & Snyder, M. (1997) *Fortress America: Gated Communities in the United States* (Washington DC: Brookings Institution).

Bonanza Village Homeowner Association (1983) Letter to City Attorney George Ogilvie, 22 August.

Bonanza Village Homeowners' Association (1997a) Letter to Bonanza Village residents, January.

Bonanza Village Homeowners' Association (1997b) Block Wall Update, April.

Bonanza Village Times (1998) Special Notice, 25 February.

Caldeira, T. (2000) *City of Walls: Crime, Segregation, and Citizenship in Sao Paulo* (Berkeley, CA: University of California).

CityLife (2000) Private government: Mayor jumps into fray over homeowners associations, 30 March.

Community Associations Institute (2004) Data on US community associations, Available at http://www.caionline.org/about/facts.cfm.

Couzens, F. (2003) Bonanza Village owner gets TRO in SID payment dispute, *Las Vegas Tribune* (on-line edition) 8 August.

Edwards, J. (2000) Building fences: controversy stirs an enduring neighborhood, *Las Vegas Review Journal*, 9 July.

Gottdeiner, M., Collins, C. & Dickens, D. (1999) *Las Vegas: The Social Production of an All-American City* (Oxford: Blackwell).

Hawley, C. (1982) Memorandum re: citizens' complaint, Bonanza Village street vacations, from City Clerk of Las Vegas, 7 June.

Las Vegas City Council (1998) City Council Minutes, 28 September.

McKenzie, E. (1998a) Homeowner associations and California politics: an exploratory analysis, *Urban Affairs Review*, September, 34, pp. 52–75.

McKenzie, E. (1998b) Reinventing common interest developments: Reflections on a policy role for the judiciary, *The John Marshall Law Review*, Winter, 31, pp. 397–427.

Moller, J. (2001) Council delays decision: fate of shelter will be decided in two weeks, *Las Vegas Review Journal*, online edition 19 April. Available online at http://www.reviewjournal.com/ivrj_home/2001/Apr-19-Thu-2001/news/15904367.html (accessed 22 January 2005).

Null, H. (1981) Letter to Bonanza Village Homeowners Association from Chief of Planning of City of Las Vegas, 14 September.

Ogilvie, G. (1982) Letter regarding Bonanza Village street vacations to City Attorney of Las Vegas, 7 May.

Silver, K. (2003) Isn't that special? Special improvement districts not always true to their name, *Las Vegas Weekly*, 1 May.

Thompson, M. (2000) Letter to Christine Monroe re: Bonanza Village Security Wall, 14 August.

Wills, L. (2000) Trouble in paradise: residents of a pastoral old Las Vegas neighborhood are fighting over an expensive plan to keep out the riffraff, *CityLife*, 13 July.

Zapler, M. (1998) Neighborhood growth measure OK'd, *Las Vegas Review Journal*, 10 March.

Homeowners Associations, Collective Action and the Costs of Private Governance

SIMON C.Y. CHEN & CHRIS J. WEBSTER
School of City and Regional Planning, University of Wales Cardiff, Cardiff, Wales, UK

(Received November 2003; revised June 2004)

KEY WORDS: Collective action, homeowners association, rent-seeking, gated communities, transaction costs, institutional evolution, neighbourhoods, Taiwan

Introduction

Apartments and condominiums lead to complicated relationships between owners, residents, developers and managers. They demand higher levels of organised co-ordination and co-operation than traditional neighbourhoods and their rise in popularity in recent years has therefore been associated with innovation in urban governance institutions. Asian cities are fascinating laboratories within which to observe the co-evolution of new modes of development and new modes of governance. The urban hubs of Taiwan's economic success have had to adapt, discover and invent ways of accommodating phenomenal growth and structural change. Privately governed

Table 1. The number of registered HOAs in Taipei 1995 to 2000

Year	Number of registered HOAs	Number of new condominium communities	Ratio (%)
1995	107	606	17.66
1996	147	507	28.99
1997	162	537	30.16
1998	176	514	34.24
1999	174	472	36.86
2000	82	403	20.34
Total	848	3039	22.57

Source: Lin, 2001, pp. 1–6.

neighbourhoods have burgeoned as in other Asian cities but the path of institutional evolution is shaped by unique local factors as well as global influences. An important development in this respect is the Condominium Management Law (CML) of 1995, designed to attend to the practical needs of homeowners living under the discipline of co-ownership neighbourhood contracts.

However, even the best *ex ante* legislative designs fall foul of unforeseen behaviour, and since the enactment of the CML the rate of the HOAs' registration has been disappointingly low. The problem is indicated in the statistics for registered HOAs in Taipei, shown in Table 1. From 1995 to the end of 2000, there were 686 registered HOAs in Taipei, representing 22.57 per cent of all condominium apartment schemes and landed co-ownership schemes (denoted 'communities' in the Table) constructed in that period and just 4.6 per cent of the total number in Taipei. The percentage in other local government jurisdictions is likely to be no more than 5 per cent (Lin, 2001).

One reason for the problem is suggested by another survey that indicated that HOA formation becomes less likely as the number of residents increases. Table 2 presents the results of Guo's 1999 survey, in which he collected the numbers of registered HOAs in three major cities, Taipei, Taichung and Kaohsiung from 1996 to 1998, and divided them into three different sized groups. 'Large communities', in Guo's study, contain over 201 units; middle-sized communities, between 101 and 200 units; and small communities, no more than 100 units. He found that the larger the size of community the lower the rate of successfully establishing HOAs. Between 64 per cent and 72 per cent of small communities formed HOAs in these years. The range for middle-sized communities was 19 per cent to 24 per cent; and for large communities, 8 per cent to 12 per cent.

Table 2. Registered HOAs in different sized communities 1996–1998

Size	1996	1997	1998	Total
Large community Over 201units	17 (12.23%)	69 (8.00%)	58 (8.43%)	144 (8.53%)
Middle-sized community Between 101–200 units	33 (23.74%)	211 (24.48%)	134 (19.48%)	378 (22.38%)
Small-sized community Below 100 units	89 (64.03%)	582 (67.52%)	496 (72.09%)	1167 (69.09%)
Total	139 (100%)	862 (100%)	688 (100%)	1689 (100%)

Source: Guo, 1999, p. 61.

He concluded that larger communities find it more difficult to reach the legal requirements necessary for HOA formation. A Hong Kong study (Chan, 2002; Lai & Chan, 2004) repeats this finding, showing a clear negative relationship between numbers of residents and success of group action by collective decision. Property owners in larger private housing estates in Hong Kong showed a greater reluctance to form corporations to replace developers' management committees.

Factors other than community size also influence the HOA formation decision, for example, building type, cultural acceptance of private responsibility in community management, strength of enabling legislation, strength of municipal government, and neighbourhood homogeneity. The Taiwanese and Hong Kong studies to some extent control for these and other influencing factors and it may be presumed that they successfully capture the numbers effect.

The unpopularity of HOAs in Taiwan, or indeed in any other country where they are slow to form, therefore might not be due to a lack of any perceived net benefit, but to the size of expected individual net benefit and related problems of free-riding. This is Olson's (1965) problem of the *latent group*: the expected gain to any one individual is insufficient to trigger joint action, even though there are overall net gains to acting jointly. See also Stevens (1993) for an analysis of collective choice problems, including a discussion of Homeowners Associations.

Collective action problems are not confined to the difficulty of group formation. Once established, residents have to be given incentives to participate in on-going governance. Numbers, might be assumed a priori, to be a problem here too, as will the size of anticipated benefits and costs of participation. The costs in question, are largely the costs of transacting, or more generally, co-operation costs and these can be reduced by institutional design (Webster & Lai, 2003). A survey of Californian residents' associations by Barton & Silverman (1987) indicated that 23 per cent of associations had difficulty filling seats on the board. In 19 per cent of associations, one board member did all the work; less than 1 per cent of member residents had ever served on a committee or the board itself; and only 11 per cent of members had contributed to the association in a voluntary capacity. The same problem is manifest in Taiwan. Wang *et al.*'s (1993) survey of HOAs before the enactment of the CML showed that most residents were not keen to participate in board-level management. Lin's (2001) investigation of office buildings in Taipei similarly found very few owners wanting to participate in the HOA-governed commercial developments. Chin (2002) reports that some communities have resorted to 'draws' to 'elect' their board members.

Aggregate patterns like these reveal the self-interested nature of people and their passive attitude toward participation in public affairs. This understanding is not fully recognised in the institutional design of Taiwan's CML, which has the underlying intention of encouraging group involvement and self-governance. The design of voting rules and the liabilities of board members clearly make assumptions about individuals' willingness to participate that are not sustained in practice, as will be elaborated later in the paper.

Notwithstanding the latent group problem, HOAs have been forming, albeit at slower rates than hoped for, under the influence of another group that has clearer incentives to organise collective action. The legal requirement for every condominium and co-ownership community to have its own HOA creates a potentially substantial market for entrepreneurial organised governance and there are a large number of latent HOAs with significant pools of public funds waiting to be exploited. As a result, the number of property management companies (PMCs) has increased rapidly in recent years. According to Taiwan's Construction & Planning Administration statistics, the number of PMCs has reached 412 and continues to rise (CPA, 2003). This sector is playing an important role in the development of HOAs in Taiwan and can be seen as a market response to the type of collective action problem identified by Olson.

The remainder of the paper is organised as follows. The next section develops a theoretical framework for analysing collective action and rent-seeking problems in HOAs. The following section describes the institutional development and design of HOAs in Taiwan. The fourth section describes the changing nature of the property management sector in Taiwan. This is followed by a discussion of property management companies and developers in the context of group formation and rent-seeking problems in HOAs. The final section presents some conclusions.

The Logic of Collective Action and Rent-Seeking

Olson's (1965) seminal work on collective action pointed out a fallacy in traditional group theories. They assumed that if a group had some reasons or incentives to pursue its members' interests, then rational individuals in that group would also have the incentive to support the collective endeavour. This is logically fallacious, especially in large groups. The larger the number of people belonging to an interest group, the less likely is the successful formation and sustained management of the organisation. First, this is because the larger the number of members in the group, the greater the organisation costs, and thus the higher the hurdle that must be jumped before collective net benefit can be realised. Large groups have high set-up costs, including the costs of co-ordinating 'buy-in'. Second, many of the benefits of group-oriented action are, by nature, non-excludable and indivisible. They are local public goods and each individual may be better off if he or she free-rides on the group participation of others, reaping the benefits without incurring the costs.

The larger the group, the higher the organisation costs and, for a given level of total benefit, the smaller the shares of the total benefit that will be going to any individual, including the individuals who assume the costly task of organisation and leadership. Olson's question is one of incentives to bear the costs of the collective action. In small group situations, the collective action problem may be more easily overcome. If the personal benefits derived from group organisation are highly concentrated in a small number of people, then the higher per capita stake may outweigh the costs of participation. This is the case where a smaller group of activists organise a large group, having the incentive of a greater than average share of benefits secured by their organising position of power. It is also generally easier to monitor and punish free-riding behaviour in small groups.

These relationships were also addressed by the Nobel economist James Buchanan in his 1965 statement of the economic theory of clubs. This challenged the conventional welfare economic distinction between public and private goods by recognising that few goods are in infinite supply and that the benefits and costs of collectively consumed goods vary with numbers of co-consumers. His analysis focused on the optimal size of clubs but it can also be taken as a sophisticated model of group formation and sustainability since formation and sustainability must be a function of a club's attractiveness.

The attractiveness of a club, an organisation that supplies local public goods to a membership of finite size greater than 1, is a function of the costs of supplying the shared 'club good(s)' and the benefit created. Total costs and benefits per person of some good or service of fixed quantity vary with the number of people jointly consuming and paying for it. Benefits from a children's play area, for example, might be expected to rise initially as more join the club until such time as congestion sets in, when they start to fall. Hence a benefit curve, if drawn, would be concave. Average costs, on the other hand, start off high with small numbers (spread over just one household in the extreme) and fall off as additional members join. At the point where distance between average benefit and costs is greatest, maximum net benefit is derived for club members and this defines the optimal membership given the fixed quantity of good. There will be a different set of cost-benefit relationships (curves) for each quantity of the good, however and this leads to a set of optimal membership sizes for each size of facility. The benefits and costs from larger facilities (an expanded children's play area, for example) are greater but their capacity is greater and optimal size of consumption-sharing group is larger. Buchanan showed that there is another way of looking at the problem of optimal size, starting with a fixed number of members and varying the quantity of the club good. As the size of a facility increases for a fixed membership, benefits increase at a decreasing rate reflecting diminishing returns, either flattening out or down-turning if the surplus actually reduces a household's utility (the playground gets too big to feel safe within). Costs rise with the size. For each club size there is an optimal quantity given by the maximum distance between cost and benefits. Buchanan's contribution was to show that, in theory, there is a uniquely efficient club size given by plotting optimal community size as it changes with facility size and optimal facility size as it changes with community size.

In the context of group formation, Buchanan's analysis shows that groups of any size can efficiently allocate costs and benefits to members, so long as the number of members and quantity of club goods can be co-ordinated. This is possible with entrepreneurial clubs that ration consumption by membership fee. In this case, membership overcomes the free-riding problem and the fee efficiently allocates costs and benefits. Buchanan's club theory can be applied to non-entrepreneurial urban clubs (see for example, Heikkila, 1996; Webster, 2002, 2003; Webster & Lai, 2003) but should in that case be taken as complementary to Olson's analysis. Buchanan demonstrates the possibility of efficient delivery of local public goods via clubs (as does Charles Tiebout, 1956, in the context of cities, conceived of as competing clubs); but Olson demonstrates the problem of club formation (in the absence of effective co-ordination that yields efficient pricing and prevents benefit leakage). In one sense, therefore, the issue of group formation and optimal size is an institutional one: what institutional and organisational order can effectively deliver the benefits of collective consumption (see also North, 1990; Williamson, 1998)?

Olson's and Buchanan's models both articulate ideas about the costs and benefits of co-operating under unified management. As such, they have an interesting parallel in modern theories of the firm and, more generally, institutional economic theory developed from Ronald Coase's foundational inquiry into the nature of the firm (Coase, 1937). He asked why firms exist and suggested it was because transacting for all resource and knowledge exchanges in the open market is too costly. The costs of searching for exchange partners and making and policing contracts is reduced by combining rights within a unitary legal trading entity. Asking the logically subsequent question: why is the economy not one single firm?, Coase also pointed to co-operation (transaction) costs. The costs of organising co-operation through hierarchical management within a firm eventually become too large and it is cheaper, at some point to contract out. Hence the notion of optimal firm size being determined at the margin by the relative value of internal organisational costs and external market transaction costs.

In the context of governing and managing urban neighbourhoods, the costs of co-operating over shared resources, local public goods and externalities via individually-negotiated agreements is excessive. It is more efficient to pool resources and rights either through traditional municipal government or through co-ownership tenure and community or private governance. However, the costs of organising and sustaining collective governance grow with the size of the organisation (firm, HOA, government). At some point, the size of the organisation becomes too large: organisational costs exceed individual expected gains compared to the alternatives. Hence, different types of organisation may succeed each other over time through a process of institutional competition.

Olson emphasised the incentive problem with large numbers but the more general problem is twofold: information and opportunism. Because information is not perfect, the co-ordination of large numbers becomes prohibitively costly (the organisation problem); the true collective and pro rata net benefits of collective action are unknown (the incentive problem); and the task of monitoring and policing contributions to collective action easily becomes overwhelming (the free-rider problem). In the light of the information problem, individuals find opportunities to capture shared (public domain) resources for their own benefit. Policing this kind of activity becomes another organisational liability that adds to the cost of large organisations.

The flip-side of Olson's logic of collective action and inaction on the demand side, therefore, is the logic of rent-seeking, or the capture of the political market by special interest groups (seeking to expropriate economic 'rent', value, by opportunistic behaviour). Rent-seeking behaviour by a special interest group is that which generates substantial personal benefits for a limited, identifiable number of constituents, while imposing a small individual cost on a large number of unidentified members of the public. The gains to members of the rent-seeking group may be less than the total loss to the larger 'public' but the interests of the latter may not be represented because losses are thinly dispersed across an unidentified public, none of whom may have sufficient incentive to mobilise (Olson, 1965, 1982, 2000; Pennington, 2000). The rent-seeking problem may be compounded when voters have to monitor the actions of politicians and public sector bureaucrats. Given the vast number of interventions by government into economic and social life, the costs of acquiring accurate data must be extremely high and the public has little chance of monitoring the degree to which they are being exploited by various interest groups (Tullock, 1989, 1993).

Taiwan's record of HOA registrations and residents' passive attitudes may be understood in the light of these ideas about information imperfection, co-operation and organisational costs. However, the collective action problem in larger groups is not necessarily insurmountable. Olson suggests four strategies: privileged group, federal group, coercion, and positive inducement. A privileged group is a group in which members, or at least some or one of them, have an incentive to see that the collective good is provided, even at the expense of bearing the full cost. A federal group is one of a number of smaller groups, each of whom has a reason to join a federation representing the large group as a whole. Coercive incentives induce collective action by threat of punishment. Positive inducements give incentives to members of a latent group to become active, with benefits greater than the perceived costs, with inducements being either material or non-material.

Taiwan's Condominium Management Law addresses the collective action problem principally through the third of these: coercion with threat of legal action. It is the fourth, however, that provides a more spontaneous momentum in populating the institutional space created by the law. Property companies have found in the law, an opportunity for business that gave them incentives as organisers of joint action and suppliers of collective goods in a 'club' market. This relates to another powerful idea from institutional economics: that it is the owners of the scarcest factors of production that tend to organise co-operative activity, for example, form firms, organise governance and build cities (see Barzel, 1997). Compared to residents, property firms have superior knowledge about how to turn neighbourhood collective goods problems into profit. This includes knowledge about alternative management solutions, access to government, access to capital and so on. Put another way, they can organise at lower cost and with lower risk. However, by taking a lead role, they put themselves at an information advantage compared to residents, especially, following Olson, in larger HOAs, and this increases the chance of opportunism and rent-seeking. The paper develops these ideas further and applies them to Taiwan, but first there is an account of the development of Taiwan's HOAs.

The Institution of HOAs in Taiwan

Primitive HOAs and their constitutions existed before the Condominium Management Law (CML) of 1995 and like that law, attempted to attend to the practical needs of homeowners. In some developments, developers helped homeowners create their own association; in others, residents felt the need to establish an organisation and rules for managing common property (Wang *et al.*, 1993).

The first report to explore problems in common property management in Taiwan was the 'Proposal for Management and Maintenance in High-Rise Buildings and Residential Communities' prepared by the China Credit Information Service Institute (CCISI) in 1984. This report noted that it is difficult to maintain and manage communal issues and to enforce the constitution and rules of HOAs without government backing and clarification of their legal position. Problems of enforcement and legal status impeded HOAs' ability to mediate in conflicts between owners and forcibly levy service charges for common property management. With an unclear assignment of property rights over shared goods and liabilities, residents' motivations to participate in communal affairs and to obey rules were critically weak, leading to deterioration of environmental quality and other neighbourhood attributes. The Council for Economic Planning & Development also expressed a similar view in the report, 'Housing Policy in Taiwan' in 1988, arguing that without a new Condominium Management Law, it would be impossible to coercively direct homeowners' behaviour with regard to the use of common property or to enforce HOA rules. Without defining rights, liabilities and associated sanctions more clearly, conflicts would increase and a decline in the living environment and public facilities would ensue (Young, 1991).

Both reports urged government to enact a Condominium Management Law in order to encourage the development of HOAs in high-rise buildings and other residential communities, as well as confirm their legal power to enforce their constitutions for the maintenance of common properties. The CML was completed and legally enacted in April 1995, giving HOAs a strengthened legal status and authority to enforce communal constitutions.

The CML requires any newly constructed apartment condominium or landed community involving shared or fractional ownership to establish a HOA when two-thirds of the total common property-owning units in a new construction have been constructed. Developers should call all owners to participate in a meeting, the 'Sharing Owners Meeting', to establish a HOA by electing a board of members and passing a constitution. Developers are also required to finance an initial public fund, a responsibility that lasts until the community has been established for one year. Although the CML requires developers to initiate HOAs in newly constructed communities it does not impose sanctions on those developers that fail to do so. The law provides even less purchase on older communities constructed before the enactment of CML.

In these pre-1995 developments, it has proven very difficult without the aid of developers, to obtain the required two-thirds attendance at an initial meeting to establish HOAs. Obtaining the agreement of three-quarters of those in attendance has also proven problematic. For example, in Taipei City, there are 40 000 apartments that were constructed before 1995 and most of these have yet to initiate a HOA (*Mingsang News*, 2002). In the light of this problem Taiwan's government issued a modification to the CML in 2002, which gives local governments greater coercive power to initiate a HOA (Lin, 2002). This empowers government to take responsibility in older communities that cannot form their own HOAs. For a discussion of government's role in reducing transaction costs in communal management through law, see Walters & Kent (2000).

The HOA institution constructed by the CML includes several elements that make it similar to public government. These, it is suggested, may also build in certain government-like collective action problems. HOAs are dissimilar to public democratic government in that voting rights are based on share of common property owned. Owners of more shares have more than one vote. In other respects, however, the 'Sharing Owners Meeting' resembles a general election in democratic countries. In particular, the board, the supreme authority in the community, is reshuffled by vote at least once a year. During the meeting the board is elected and decisions are taken about communal issues. For important decisions (such as initiation of the HOA) there needs to be a two-thirds attendance of all sharing owners and an agreement of three-quarters of those in attendance. Important decisions also include modification of the constitution, large projects of communal construction and improvement, and coercive expulsion of residents who have violated communal rules. Other decisions only need to have a quorum of half the sharing owners and the agreement of half of these (Articles 29 and 31, CML).

Another democratic feature of the CML's model of governance is the extension of certain rights to non-owners. HOAs consist of units of 'residents' not only homeowners. In other words, tenants can take part in a HOA and even become board members. They have a right to participate in HOA meetings apart from the Sharing Owners Meeting, and through their influence in these and in office on the board, have a chance to direct the day-to-day allocation of resources within the organisation. This seems to be in contrast to American homeowners' associations where non-owning residents are more thoroughly excluded from real engagement in community affairs (McKenzie, 1994). The rights assigned to residents seem to build in the potential for internal political problems similar to conventional government. A significant number of non-owning residents pay HOA fees, which are passed on to them by owners. If enough non-owning residents are on the board, they also get to manage the expenditure of the public fund. In general, non-owning residents have an incentive to spend the fund prudently since the costs fall to them.

They may be assumed to have a shorter-term interest than owners and if prudence or short-term interest compromises the interests of owners (largely, investment value) then competition between owners and renters may introduce political complexity. In principle, owners can organise an Owners Meeting and seek to regain control of the board, perhaps firing its members. However, it might be supposed that non-owning residents in positions of power on boards use their information advantage to protect their own interests. A 'moral hazard' problem exists in that it is board officials who organise meetings and collect and process management information. With non-owning residents' interests represented at the heart of the day-to-day operations, any such disputes are unlikely to be resolved without considerable organisational and transaction costs. This may well deter owners from fighting certain issues and will effectively leave many resources in the public domain where they can be captured by non-owning residents. Thus the political economy of municipal government emerges in its own version within HOAs.

Board members, headed by the president and vice president, are charged to fulfill the HOA's mission. They have responsibility for maintenance of common areas and management of association assets, which can range from very little to millions of pounds. They are responsible for collecting monthly service charges and special assessments for particular purposes and for enforcing the communal constitution and rules. Board member job descriptions are defined in the CML with reference to government style public service ethics. They are unpaid and are expected to be obedient servants, serving the public interest. If they fail to serve in an appropriate manner they are liable to the Sharing Owners for their errors. Presidents have significant legal responsibility, representing HOAs as defendant or plaintiff in any lawsuits.

Given the lack of positive incentive for either residents or owners to seek board membership and the liabilities of time, personal costs, risked reputation and risk of legal action, it is not surprising that few HOAs have been initiated unilaterally by owners without the aid of a third party. There is now a look at the third-party involvement of property companies.

The Market Development of Property Management

The growth of property management companies in Taiwan has paralleled the growth of common-ownership developments. In the early days, 20 years ago, companies of this type only provided services such as cleaning and security. Today, as a result of the enactment of the CML and increasing numbers of HOAs, property management companies have become a multi-product industry providing an increasing range of private governance functions.

According to Lin's (2002) analysis of this market, modern property management companies are spin-offs from or expansions of real estate companies, property management departments of larger companies, upgraded cleaning and security companies, as well as multinational firms that developed their specialism elsewhere. Conventionally, each company in this highly competitive market (with easy entry) has occupied only a very small market share. Each company needs its niche in order to survive and companies have therefore focused on market segmentation—creating specialist markets in the management of shopping malls, residential communities, commercial buildings, and in different styles and sizes of these.

Competition forces firms to discover how to provide more and better services at cheaper prices. Property management companies must now provide more than just security and the traditional maintenance of the physical environment and common facilities. They must be able to supply personal services, for example, calling taxis, ordering flight and theatre tickets from a communal service counter, as in hotels, and consulting services, including legal advice related to investment in real estate and the letting of houses. Recently, perhaps as a result of increasing consumer differentiation between commoditised neighbourhoods, emphasis has been placed on developing communal spirit and enhancing 'atmosphere' and cultural assets through communal activities such as community festivals, events and communal courses such as painting and handicrafts.

Companies have used technology to develop new and refined products, setting up community websites and providing community management software so that residents can remain linked through remote control.

In recent years, a spate of serious natural disasters have plagued Taiwan including earthquakes, landslides and floods and some companies have endeavoured to provide hazard mitigation plans as an additional selling point (Lin, 2002).

Property companies have been resourceful in cutting costs as well as improving services. Many companies have adopted the ISO 9000 system to standardise forms and operational processes, another sign that competition in the sector drives product quality. Many companies have also introduced Management Information Systems to computerise working environments and Enterprise Resources Planning to integrate and reorganise associated enterprise resources. Faced with widely dispersed employees and resources, multiple products and services and many different kinds of maintenance tasks, i.e. a big information and organisation problem, some companies have developed remote monitoring systems that computerise stocks and tasks in each community and facility and use GPS to monitor employee movements (Yeh, 2003).

Illustrating the maturing of this private governance market, private professional associations have started to emerge. The Building Manager Association (BMA) was formed in 1997, the most influential private association in the development of HOAs. It conducts government-recognised training courses in property management and undertakes research for the government and property management practitioners into the effect of new policies, lawsuits and proposals for modifications to the CML. The first modifications to the CML were proposed by BMA in 1999. Competition is provided by a second similar association, the Taipei Building Community Service Association founded in 1999.

Thus, through market competition, property management companies appear to be increasingly well placed to organise efficient community management and to service private community governance. As described in the next section, they have also used their increasingly specialised knowledge to lead in the organisation of new HOAs.

The Role of Property Management Companies and Developers in HOAs

Property management companies and developers have an important influence in the initiation and operation of HOAs in Taiwan. Developers are required by law to help organise associations in new developments. Property management companies and associated private professional associations voluntarily help existing communities to organise collective action. From 73 randomly selected HOAs, Guo (1999) found that almost 70 per cent of HOA initiations were aided by property management companies, 10 per cent were helped by developers and around 15 per cent were established by residents themselves.

While the Taiwanese government is considering lowering the threshold for agreement among residents to relieve the collective action problem, property management practitioners have developed a more sophisticated stratagem. Guo (1999) found that most property management practitioners do not take the legal requirements for establishing a HOA too seriously. In their experience, the problem can be resolved by holding many meetings in small groups, for example, dividing residents based on different areas, buildings and flats. In these smaller groupings it is easier to convince owners and residents of the benefit of creating a HOA. This also makes it easier to obtain the legal quorum and consensus at subsequent homeowners' meetings. This is rather like Olson's 'federal group' strategy, dividing a group into a number of small groups, each of which has a reason to join with the others to form a federation representing the large group as a whole.

Property companies are motivated to help communities initiate HOAs by the benefits of enlarging the market, seeking out niches, and building favourable client-contractor relationships. Shiu (2001) reports a situation in Taoyuan County, Taiwan, where HOAs are developing very quickly. A few may be due to civil-minded residents but more are due to the promise of huge public funds available after the HOAs have been initiated. Owners in a development get advice from the property company and as a result they face lower set-up and other organisational costs and generally, lower *ex ante* contracting costs. The property company gets a contracting partner with a stronger legal standing, thus reducing *ex post* contract monitoring and policing costs. The relationship developed in the process eases the establishment of subsequent operational contracts, and the property company gets preferential access to the newly created demand for its increasingly sophisticated range of services.

Property management companies and developers have therefore contributed to the efficient development of HOAs. This has worked because their interests coincide with those of the owners and residents in the community: property companies know how to deliver efficient community services; 'latent group' communities have potential demand and, it seems, can be convinced of the net benefits of collective action if someone else assumes liability for the transaction costs.

It is worth considering the conflicts of interest that might risk distorting this otherwise happy exchange. McKenzie (1994) has questioned why there are so many lawsuits in American HOAs. Many cases show that the attorney who advises the board on whether to file a suit will handle the litigation and receive substantial fees. This raises the question of whether legal advice in these situations is as neutral as it should be. Yip & Forrest (2002) also observed similar conditions in Hong Kong, where condominium owners were not satisfied when a large property management company displayed monopolistic behaviour. Conflicts emerged particularly in relation to the level of service charges and the standards of repair and maintenance. This is directly analogous to problems that typically emerge in publicly managed rental housing, where public agencies have an effective monopoly on social housing provision.

An investigation in NangKang Software Industrial Park in Taiwan by one of the authors of this paper, found that the developer reaped a monopoly profit through owning the dominant power in the HOA. Homeowners were at risk of exploitation, having to take a contract at above market prices with the developer's subsidiary firm to maintain their environment.

More generally, and extending this to property management companies, there is a fairly high degree of asset specificity in a contract between a management company and a HOA. Following Williamson's (1985) transaction costs theory and Grossman & Hart's (1986) property rights theory, Deng (2003) shows how the specificity of knowledge and relationships in contracts over the provision of neighbourhood services introduces inefficiencies in contracts between provider and consumer. He demonstrates the efficiency of a variety of contractual arrangements in the provision of community services, bearing in mind the opportunity that suppliers of specific assets and services have for rent-seeking by holding-up clients or customers in negotiating for better terms.

Rent-seeking Inside HOAs

Holding-up is one way of acting opportunistically. Another is using the political system for individual advantage. Resource allocation in HOAs is by voting rules and this affords scope for opportunist behaviour. Benefits enjoyed by owners of units in a condominium are likely to be a function of number and characteristics of residents, as well as the heterogeneity in the attributes and value of individual units. Deviations from proportionality between assessment and benefits mean that some people's property rights are left in the public domain and can be appropriated by others. Decision making by majority vote can allow the majority to gain at the expense of minorities. For example, if votes are proportional to the value of units but assessments are uniform, owners of more valuable units can exploit the voting process to reap gains at the expense of owners of smaller units (Barzel & Sass, 1990). Rent-seeking by the majority can be found in Taiwan. 'Majority violence' has been reported in the form of unreasonable service charges, for example, ground floor owners having to pay the same service charges as the higher floor owners although they rarely use lifts and residents having to pay fees for parking lots whether or not they have cars. Practices such as these led the Taiwanese government to modify the CML to allow minorities to challenge HOA decisions on service charges through arbitration and the courts (Cheng, 2002).

A degree of redistribution is inevitable in any fiscal system, whether it is government tax or HOA assessment. However, the contractual nature of HOA governance offers greater scope for calibrating fee regimes more efficiently. The scope for matching fees to benefits is limited by the cost of legal challenges and the cost of administering a differentiated fee system. At some point, the cost of clarifying property rights exceeds the benefits of further refinement.

The public fund created by member fees, with the initial help of developers, is used for contracting property management companies, hiring security guards, repairing common facilities and is an incentive for rent-seeking—the more so the larger the fund. Conflicts over, and abuse of, public funds can be expected to be more common in larger communities where the funds are sizeable. For example, 'Lifetime Town' in Taoyuan County, contains 2369 units and is a typical huge gated community with a considerable fund. When the project was completed in 1999, the developer handed over £1.2 million to the public fund. In 2001, two HOAs existed simultaneously in the community because the existing board members refused to hand over power to the new elected board members and both declared they were the legitimate representative association and the other illegally established. This situation arose as a result of the two groups' competing proposals for using the communal public fund and it led to political and gang rivalry for the prize of

HOA control in the town (Ro, 2001). The conflict was finally resolved by the courts but not without the dissipation of significant costs in terms of time, private and public funds and communal division. The conflicts still continue in the community—the community's website tells a story of rumours, lawsuits and ill feelings that have apparently resulted in many disappointed residents moving out.

This seems to suggest an apparent twist to Olson's basic story. The larger the HOAs, the greater incentive for rent-seeking elements within latent groups to engage in collective action. It is not a twist, of course, since Olson's free-riders desist from action because of low expected individual gains compared to engagement costs. Individuals, groups and firms with the knowledge to secure disproportional returns from public domain resources view the calculus positively and pro-actively engage. The dynamic is the same as in public government, where single interest groups often find it worth their while to engage in public participation exercises.

HOAs therefore face the same set of collective action problems as governments, more so to the extent that community residents under-invest in gathering information to monitor the operation of their HOA. In public government it is often the press who has the biggest incentive to monitor rent-seeking.

Conclusion

Collective action problems are inherent in the formation and running of HOAs. Owners, residents, property management companies and developers each have different incentives to influence the creation and operation of HOAs. Developers who divest their property interest in a development when all units are sold, have no material interest in creating a HOA even though they are the most able to do so. With no title to the property after the last unit is sold they have no residual claim on the resources that stand to be enhanced by a HOA. However, during the period of marketing and sales they have an incentive to bequeath an organisation that promises to maintain and enhance asset value. The latter incentive was apparently not strong enough for widespread developer initiation of HOAs in Taiwan and it had to be backed by government coercion under the Condominium Management Law. Property management companies have a more obvious material interest in HOAs, however, and are actively touting for business among the many co-ownership developments in Taiwan that still do not have a HOA.

The voting system in HOAs is not like those in stock corporations. If voting is not well designed, some property rights remain ambiguously allocated and will encourage rent-seeking behaviour by those able to take advantage. Larger communities with substantial public funds incur higher monitoring costs and this further encourages opportunistic behaviour. Although voting is also used in private firms to allocate resources, it functions differently. Barzel & Sass (1990) make the point that in a stock corporation, returns to shareholders are uniform across shares. Shareholders are given votes in proportion to share holdings and additional assessments are zero. Therefore, any project that demonstrably enhances the value of the corporation is likely to receive a unanimous vote. Property rights are delineated well under this voting system. Voting in HOAs does not perform such a clear cut job in assigning property rights unambiguously. It lies somewhere between that of private firms and government. In Taiwan, with the Condominium Management Law's mixture of democratic and corporate features, it looks more like a public entity than a private firm.

The fact that over a quarter of a million Americans now live in developments governed by a HOA must say something about the efficiency of contractual government. HOAs, condominium apartments, malls, co-operatives and other forms of club community demonstrate that civic goods and services can be supplied as tie-ins with private goods and are financed by ground rent, membership fees, room rent and special assessments (Foldvary, 1994; Nelson, 2002; Pennington, 2000; Webster, 2001, 2002). However, evidence documenting the organisational costs of contractual governance suggests that these may not necessarily be low. The paper has discussed some of the inherent collective action and rent-seeking problems in Taiwan's HOAs. McKenzie's work on American HOAs is important in suggesting that transaction costs of contractual government may not be lower than conventional government. Indeed, they may be higher. More clearly defined property rights—delineated in covenants, conditions and regulations—paradoxically generate a finer grain of monitoring, enforcement and dispute costs. The crucial question that must be addressed jointly by the private neighbourhood industry and government is how to lower these costs. There are plenty of examples of the way institutions are evolving in this respect, including recent state laws in the USA designed to keep proliferating HOA litigation out of the public courts (McKenzie, 2003b). This effectively forces the HOA industry to internalise its own organisational costs through the use of private mediation and arbitration services.

This paper has argued that the structural problems of collective governance—information asymmetry and opportunism, free-riding and rent-seeking—are inherent in HOAs as much as in public government. HOAs in Taiwan, as elsewhere, may be creations of the market (enabled by state law) but they do not themselves create market devices for allocating collective goods. MacCallum (2002) also observed the HOA problems stemming from the political nature within HOAs' governance. He suggested that only a multi-tenant rental model (such as typically found in shopping malls, trailer parks, marinas and hotels) can resolve these problems. Where land ownership is unitary, individual unit rents act as a market signal (price) that equates the cost and benefit of collective resources and renters are free to move out or in since they have relatively low moving costs. HOAs do not do this. They create clubs and collective ownership and involve collective decision making. They permit markets to develop in club goods by defining a membership, enforcing payment and excluding non-contributors. Within these collective consumption organisations decisions still have to be made by non-market mechanisms. In addition, homeowners are not renters and cannot easily use the exit option when they are not satisfied with a HOA's performance (McKenzie, 2003a). High *ex post* moving costs mean residents are at risk of being held-up by HOAs in the same way as citizens locked into a single municipal government.

Therefore, another empirical question can be raised alongside McKenzie's: how will institutions evolve to reduce the collective action problems inherent in HOA-based contractual government? This paper has presented some insights from Taiwan's early experience, but this is a long-term question. In time it might be expected that state laws will emerge to make HOAs more efficient. This is already evident in countries around the world (Briffault, 1999; Foster, 1997; Glasze *et al.*, 2005; Malek, 2002; McKenzie, 1994, 1998; Webster *et al.*, 2002). In the longer term it might be expected that efficiency-driven changes in state and private laws, regulations and constitutions will tend to reduce transaction costs in the various competing neighbourhood management models (condominium, co-operatives, multi-tenant rental, public government and so on). For a given technology, institutions tend to drive down transaction costs and the 'true' relative advantages of the different models tend to be revealed. It is interesting, for example, to conjecture why publicly managed neighbourhoods became the dominant feature of 20th century cities; why co-operatives have not generally been popular, but why they do, however, flourish in some localities; why multi-tenant properties (sky scrapers, ocean liners, department stores, hotels) expanded rapidly in the mid-19th century and then again in the late 20th century; why condominiums and HOAs emerged as a dominant residential governance model in much of the world outside Europe in the last quarter of the 20th century.

The analysis here indicates that if HOAs are more efficient than municipal government at the present time, it may only be because, as in the case of Taiwan, its public services are delivered by property management companies which are more competitive and more efficient than direct government providers. The problems of centrally governing the allocation of scarce resources via democratically accountable decision making are generic, whether the organisation is a municipal government or a contractual and entrepreneurial club. More research is needed to address the question of whether the collective action costs are less in HOA board rooms than in town halls. They may, in fact, be higher: but then so may be the benefits delivered by decentralised and more bespoke services.

References

Barton, S. E. & Silverman, C. J. (1987) *Common Interest Homeowners' Associations Management Study* (Sacramento: California Department of Real Estate).

Barzel, Y. (1997) *The Economic Analysis of Property Rights* (Cambridge: Cambridge University Press).

Barzel, Y. & Sass, T. R. (1990) The allocation of resources by voting, *The Quarterly Journal of Economics*, 105, pp. 745–771.

Briffault, R. (1999) A government for our time? Business improvement districts and urban governance, *Columbia Law Review*, 99, pp. 365–477.

Buchanan, J. M. (1965) An economic theory of clubs, *Economica*, 29, pp. 371–384.

Chan, Y. L. P. (2002) Public participation in the management of private residential estates in Hong Kong: an economic analysis, Unpublished PhD Thesis, Department of Real Estate & Construction, University of Hong Kong.

Cheng, S. Y. (2002) Service charge will be based on the different levels of benefits, *Minsang News*, 26 February.

Chin, H. S. (2002) Can public funds invest in the Stock Market?, *United Daily News*, 1, December.

Coase, R. (1937) The nature of the firm, *Economica*, 4, NS pp. 386–405.

Construction and Planning Administration (2003) The statistic on Property Management Companies. Available at http://www.Sercpa.cpami.gov.tw/apm/dba/apmacc.accin (accessed 9 August 2003).

Deng, F. F. (2003) The rebound of private zoning: property rights and local governance in urban land use, *Environment and Planning A*, 35, pp. 133–149.

Foldvary, E. F. (1994) *Public Goods and Private Communities* (London: Edward Elgar).

Foster, K. (1997) *The Political Economy of Special Purpose Government* (Washington DC: Georgetown University Press).

Glasze, G., Webster, C. & Franz, K. (2005) *Private Neighbourhoods: Global and Local Perspectives* (London: Routledge) forthcoming.

Grossman, S. J. & Hart, O. D. (1986) The costs and benefits of ownership: a theory of vertical and lateral integration, *Journal of Political Economy*, 94, pp. 691–719.

Guo, Y. K. (1999) *A Proposal of Modifications of Condominium Management Law* (Taipei: CPA).

Heikkila, E. (1996) Are municipalities Tieboutian clubs?, *Regional Science and Urban Economics*, 26, pp. 203–226.

Lai, L. W. C. & Chan, P. Y. L. (2004) The formation of owners' corporations in Hong Kong's private housing estates: a probity analysis of Mancur Olson's group theory, *Property Management*, 22, pp. 55–68.

Lin, C. S. (2002) Market structure of Taiwanese property management, *1st Conference of Property Management* (Taipei: BMA).

Lin, C. Y. (2001) *Voluntary Environmental Management Part II: A Case Study in Wu-Ku Industry Park* (Taipei: Council of National Science).

Lin, S. L. (2002) Coercion on initiating Homeowners' Association, *China Time*, 23, May.

Lin, W. G. (2001) *Condominium Management Guidebook* (Taipei: Taipei County Government).

MacCallum, S. H. (2002) The case for land lease versus subdivision, in: D. T. Beito, P. Gordon & A. Tabarrok (Eds) *The Voluntary City* (Ann Arbor: University of Michigan Press).

Malek, M. (2002) An investigative study into club communities in Malaysia, Unpublished PhD Thesis, (UK: Cardiff University).

McKenzie, E. (1994) *Privatopia* (New Haven and London: Yale University Press).

McKenzie, E. (1998) Reinventing common interest developments: reflections on a policy role for the judiciary, *John Marshall Law Review*, 31, pp. 397–427.

McKenzie, E. (2003a) Common-interest housing in the communities of tomorrow, *Housing Policy Debate*, 14, pp. 203–234.

McKenzie, E. (2003b) Private gated communities in the American urban fabric, *Conference of Gated Communities: Building Social Division or Safer Communities?* (UK: Glasgow).

Mingsang News (2002) Few homeowners associations in older apartments, *Mingsang News*, 26 August.

Nelson, R. H. (2002) A proposal to replace zoning with private collective property rights to existing neighborhoods, in: D. T. Beito, P. Gordon & A. Tabarrok (Eds) *The Voluntary City* (Ann Arbor: University of Michigan Press).

North, D. C. (1990) *Institutions, Institutional Change and Economic Performance* (New York: Cambridge University Press).

Olson, M. (1965) *The Logic of Collective Action* (Cambridge: Harvard University Press).

Olson, M. (1982) *The Rise and Decline of Nations* (New Haven: Yale University Press).

Olson, M. (2000) *Power and Prosperity* (New York: Basic Books).

Pennington, M. (2000) *Planning and the Political Market* (London: Athlone Press).

Ro, P. (2001) Two homeowners' associations in Lifetime Town, *China Times*, 4 September.

Shiu, W. D. (2001) Huge public fund with few sanctions, *China Times*, 7 April.

Stevens, J. B. (1993) *The Economics of Collective Choice* (London: Westview).

Tiebout, C. M. (1956) A pure theory of local expenditures, *Journal of Political Economy*, 64, pp. 416–424.

Tullock, G. (1989) *The Economics of Special Privilege and Rent Seeking* (London: Kluwer Academic Press).

Tullock, G. (1993) *Rent Seeking* (London: Edward Elgar).

Walters, M. & Kent, P. (2000) Institution economics and property strata title—a survey and case study, *Journal of Property Research*, 17, pp. 221–240.

Wang, H. K., Houng, S. M., Chen, L. C. & Bi, H. D. (1993) *An Investigation on Condominium Self-Management* (Taipei: Ministry of Interior).

Webster, C. J. (2001) Gated cities of tomorrow, *Town Planning Review*, 72, pp. 149–170.

Webster, C. J. (2002) Property rights and the public realm: gates, green belts, and Gemeinschaft, *Environment and Planning B*, 29, pp. 397–412.

Webster, C. J. (2003) The nature of the neighbourhood, *Urban Studies*, 40, pp. 2591–2612.

Webster, C. J. & Lai, L. W. C. (2003) *Property Rights, Planning and Market: Managing Spontaneous Cities* (Cheltenham: Edward Elgar).

Webster, C., Glasze, G. & Frantz, K. (2002) The global spread of gated communities, *Environment and Planning B*, 29, pp. 315–320.

Williamson, O. E. (1985) *The Economic Institutions of Capitalism* (New York: Free Press).

Williamson, O. E. (1998) Transaction cost economics: how it works; where it is headed, *The Economist*, 146, pp. 23–58.

Yeh, W. L. (2003) *Strategies for How to Run a Good Property Management Company* (Taipei: CPA).

Yip, N. M. & Forrest, R. (2002) Property owning democracies? Home owner corporations in Hong Kong, *Housing Studies*, 17, pp. 703–720.

Young, Y. F. (1991) Community management and public housing ordinance, *Conference of Housing Policy and Law* (Taipei: Housing Academy).

Some Reflections on the Economic and Political Organisation of Private Neighbourhoods

GEORG GLASZE
Institute of Geography, University of Mainz, Mainz, Germany

(Received November 2003; revised May 2004)

KEY WORDS: Private neighbourhoods, urban governance, territorial club economics

The Spreading of Private Neighbourhoods: A Global Success Story?

The spreading of private and often gated neighbourhoods in many regions of the world has triggered a new and widespread discussion about the relations between social and urban development. In the USA, according to the figures of the Community Association Institute (CAI), the number of private neighbourhoods grew rapidly from around 10 000 in 1970 to more than 200 000 in 1998. As membership in the CAI is voluntary, that list probably even underestimates the number of private neighbourhoods. At the turn of the century, at least one out of six Americans is living in a private neighbourhood. Every fifth of these private neighbourhoods in the US is gated and guarded (Community Association Institute, 1999). A first analysis of the American Housing Survey 2001 reveals that 5.9 per cent of all households stated they live in a walled or fenced neighbourhood—3.4 per cent reported that their neighbourhood was access controlled by guards or electronic devices (Sanchez *et al.*, 2003). Studies in other regions of the world allow the assumption that there is also a growing trend towards private and often guarded neighbourhoods in many countries of the world (see the contributions in this volume and in Glasze *et al.*, 2005), especially in Latin America (Borsdorf *et al.*, 2002; Coy & Pöhler, 2002), Africa (de Montclos, 1997; Jürgens & Gnad, 2002; Landman, 2000) and Asia (Giroir, 2002; Leisch, 2002). On a smaller scale, some European housing markets show a tendency towards private and guarded housing estates as well, for example, in England (Blandy & Parsons, 2003), Turkey (Perouse, 2003), France (Madoré & Glasze, 2003), Portugal (Raposo, 2003) and Spain (Wehrhahn, 2003). First studies in metropolises in the former communist states even indicate a rapid increase of this kind of housing, see for example, in Moscow (Lentz & Lindner, 2003), and in Warsaw (Glasze & Pütz, 2004).

Many authors have interpreted the private neighbourhoods as a privatisation of former public spaces (e.g. Connell, 1999; Gmünder *et al.*, 2000; Judd, 1995; Lichtenberger, 1999). The value of 'public space' and its endangering through 'privatisation' became a much-quoted theme within the critique of contemporary urbanism at large (Feldtkeller, 1995; Ghorra-Gobin, 2001; Kazig *et al.*, 2003; Lichtenberger, 1999; Mitchell, 1995). The reason may be that the concept of 'public space' is related to powerful normative ideas of equal rights and political emancipation (Habermas, 1990, p. 20). However, 'public space' and 'privatisation' are extremely vague analytical categories. Therefore, it often remains unclear what exactly is privatised and how privatisation is carried out. The writings describing private neighbourhoods as a 'privatisation of public space' tend to dichotomise between a public realm and a private realm; they often focus unilaterally on material changes in space and therefore risk blocking from view a more profound and differentiated analysis of the complex socio-economic and socio-political changes which are under way with the spreading of private neighbourhoods. Therefore, after presenting a typology of private neighbourhoods in the next section, the paper proposes three analytical approaches that heuristically might be more fruitful. In order to understand the economics of private neighbourhoods, the club goods theory will be used. This approach helps us to understand their potential attractiveness for developers, housing seekers as well as local governments. However, the question remains: why do we see a spreading of private neighbourhoods especially in these days and especially in certain parts of the world? For that reason, the following section discusses the interplay between the socio-economic and socio-political changes, usually seen as effects of globalisation on the one hand and nationally or regionally differentiated governance patterns on the other hand.

In order to portray the political changes under way, private neighbourhoods are described as private residential governments in the next section. In this perspective, the spreading of private neighbourhoods might be described as the establishment of a 'new' territorial organisation on a sub-local level which enables the exclusive consumption of collective goods, and in which political decisions are taken in a kind of shareholder democracy. However, several authors have argued that the economic and political functioning of private neighbourhoods might not be as different from public municipalities as usually assumed. The last section picks up their arguments and uses the club goods approach and the idea of shareholder democracies to discuss in detail the consequences of a shift from a public to a private organisation of local and sub-local territories.

Organisational Types of Private Neighbourhoods

In traditional neighbourhoods the open spaces such as streets or parks, as well as many common facilities such as public libraries or swimming pools, are owned by public authorities and governed by local government. In the private neighbourhoods, the open spaces and the common services are managed and regulated by a self-governing organisation. In spite of differing national juridical contexts, three main organisational types of private neighbourhoods can be described which differ in the way that property rights for the open spaces and facilities as well as for individual housing units are assigned (Glasze, 2003a, p. 238; McKenzie, 1994; Treese, 1999):

- Condominiums: In addition to the individual property of their piece of land and house or their apartment, the owners hold titles to an undivided interest in the common property of streets, green spaces, facilities etc. Thus every owner automatically becomes a member of the homeowner association. The members of the association elect the board of directors, most often on an annual basis.
- Stock-co-operatives: The co-operative owns both the housing units and the common spaces and facilities. Individuals purchase a share in the entire complex. That share offers the right to use an individual unit as well as the common areas and facilities and gives voting rights in the assembly of the co-operative.
- Corporations: The common spaces and facilities are the property of the corporation. In those cases where a covenant attached to the deed for a residence or a residential lot 'automatically' makes each owner a shareholder in the corporation with voting rights according to the amount of his share, this type is often also called a homeowner association. In other cases, where the shareholders are not identical to the people owning or renting the housing units, Foldvary has labelled this form 'proprietary neighbourhoods' (1994). In a 'proprietary neighbourhood' the people living in the neighbourhood do not have any political input concerning the development of their neighbourhood. The relation to their environment may be best compared to the relation of hotel guests to the development of the hotel.

Private Neighbourhoods as Club Economies

The self-administration of private neighbourhoods provides the inhabitants with many collective goods, such as green spaces, water supply and recreational facilities. Economists have justified the fact that even in market economies some collective goods are traditionally provided by public organisations due to market failure. That is to say, the market fails to provide goods when nobody can be excluded from consumption and when there is no competition, no rivalry in consuming. Free-riders could profit from goods such as clean air or urban green spaces without paying for them. Consequently, these goods are not sufficiently provided by the private sector and the public sector has to step in. For private goods such as food there is rivalry about the consumption and third parties can be excluded from consuming. Sometimes, commons are differentiated as a third category of goods that are competed for but for which the exclusion criteria are hard to meet. Consequently, these goods often suffer from overuse, as for example, the fish population in the deep sea.

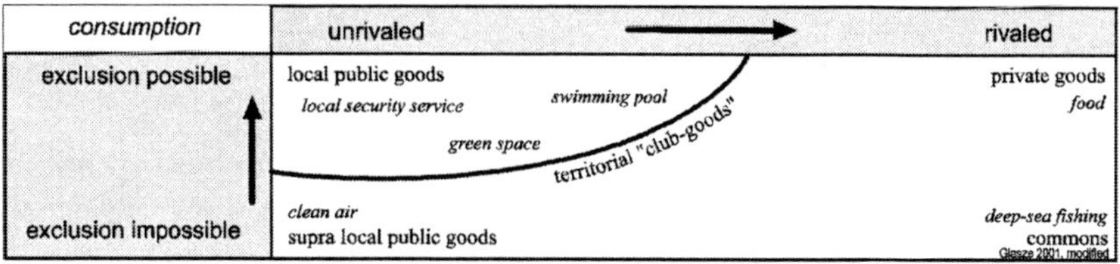

Figure 1. Private goods, public goods and club goods. Source: Glasze 2001, modified.

However, the American economist Tiebout already pointed out in 1956 that many collective goods which were generally described as public goods are 'local' public goods in the sense that they bring benefits primarily to people who stay at a certain locality. It was Foldvary who showed in 1994 that the self-administration, and as the case may be the enclosure, of private neighbourhoods solves the free-rider problem for local public goods and renders them excludable. Therefore, he and several other authors have judged the spreading of private neighbourhoods as an institutional 'innovation' which ensures a market driven and efficient supply of local public goods for the inhabitants (see for example, the contributions in: Beito *et al.*, 2002). Those who profit from the collective goods within the neighbourhood pay the respective fees (Figure 1).

Groups which collectively, but exclusively, share the consumption of specific goods on the basis of ownership-membership arrangements have been named 'clubs' and the excludable collective goods 'club goods' (Buchanan, 1965). Therefore, the establishment of private neighbourhoods with their self-governing organisation may be interpreted as the creation of club economies with territorial boundaries. The analysis of private neighbourhoods as club economies makes it possible to explain the potential attractiveness of these complexes for developers, local governments and inhabitants (Glasze, 2003c).

Developers may profit from the fact that the establishment of a neighbourhood governance structure with the power to exclude free-riders, as well as the power to regulate the use of common spaces and facilities, reduces the risk of an economic degradation of the neighbourhood. Thus, the long-term risks in investing in large-scale projects where the process of selling takes several years are reduced and the developers are able to invest more in creating and maintaining shared facilities (Weiss & Watts, 1989, p. 95). Furthermore, they can market not only the individual home but also the club goods within the neighbourhood as contractual tie-ins (Webster, 2002, p. 405). Local governments may profit from private neighbourhoods being established within their boundaries as they obtain a development which is self-financing and which may add to the local tax base. McKenzie (1994) has shown examples in the US where local governments encourage and even demand the establishment of private neighbourhoods as 'cash cows'. House seekers and inhabitants may profit from the level and the quality of local public goods supplied in private neighbourhoods. They usually offer a range of services such as maintenance, 24-hour security or waste collection as well as artificial and natural amenities such as beaches or green spaces. Furthermore, the individual owners may profit from stable home values as the self-administration assures a strict control of the social and physical environment and tries to create or keep a prestigious image of the neighbourhood.

The Global Spreading of Private Neighbourhoods and the Role of Regionally Differentiated Governance Patterns

The analysis of private neighbourhoods as club economies helps us to understand their potential attractiveness. However, the question remains: Why do we see the spreading of private neighbourhoods especially in these days and why do we see it in many parts of the world — while in other parts like for example in Scandinavia or in Germany this form of housing is almost unknown? Often, the international spreading of private neighbourhoods is vaguely related to globalisation. If one defines 'globalisation' as a 'time-space-compression', triggered through new technologies of information, communication and transport as well as the liberalisation of national and regional regulations I think it is indeed possible to identify some economic, political and social changes which render private neighbourhoods more attractive for developers, housing seekers and public organisations these days (Glasze, 2003a, p. 262) (Figure 2):

- The shift from the model of an omnipotent state to a minimal state: Particularly since the breakdown of the communist systems in Europe, ideas of deregulation and privatisation dominate the political discourse in many countries around the world.

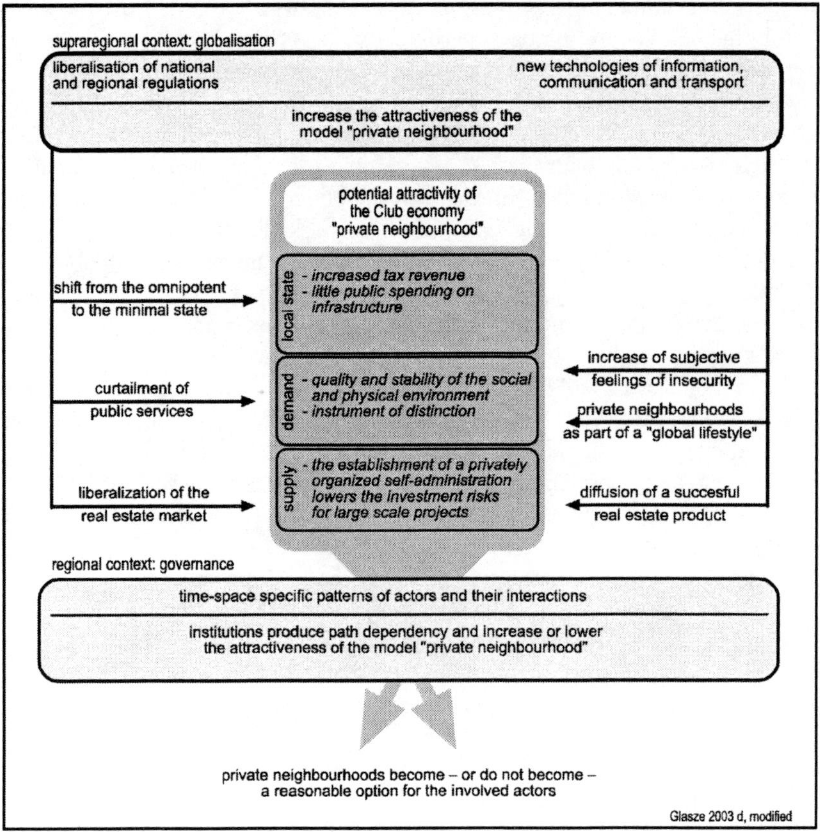

Figure 2. Globalisation, urban governance and the spreading of private neighbourhoods. Source: Glasze 2003a, modified.

The idea is to replace the steering through politics and public administration through the supposedly self-regulating forces of market mechanisms instead. Thus, Fischer & Parnreiter (2002, p. 247) show with examples in Latin America and Raposo (2003, p. 299) with examples in Portugal, that the liberalisation of real estate regulations has widened the scope for private investors and facilitated the development of private and guarded neighbourhoods. The regulation of subdivisions in Portugal has facilitated the establishment of private neighbourhoods and the liberalisation of the land market in several Latin-American countries has opened new opportunities. In a more general perspective, Graham (2000, p. 185) describes an increasing differentiation between places with an unsatisfactory (public) infrastructure and 'premium networked places' established by the private sector in many cities around the world. Empirical research in Southeast Asia (Leisch, 2002), in Latin America (Janoschka, 2002; Pöhler, 1998) as well as in the Arab World (Glasze, 2003a, p. 183) has shown that the private supply of high-quality services (such as electricity, water supply and communication services) often is an important motivation to move to a private neighbourhood. In many regions, in particular in the so-called 'Third World', the private neighbourhoods substitute public supply and regulation, however, only for a clientele with sufficient means.

- Growing feelings of insecurity: In the course of an ongoing modernisation of societies, a growing social differentiation and individualisation develop fundamental uncertainties. Informal social networks like kinship or other traditional forms of community are getting weaker and are no longer available with certainty in times of crisis. In societies which experience a rapid transformation, as in the former communist states (Glasze & Pütz, 2004; Lentz & Lindner, 2003), or in South Africa (Jürgens & Gnad, 2002), this point seems to be particularly important. At the same time, in many countries it can be observed that systems of social security are reduced. Furthermore, in many regions the establishment of competing private media leads to a growing 'scandalisation' of media coverage (Siebel, 2003). Against this background, many inhabitants of private neighbourhoods look for 'security'—the security to live in an environment whose material and social qualities are regulated by private contracts and therefore promise to be more secure than in 'traditional' neighbourhoods.
- Guarded housing estates as part of a global lifestyle: Particularly in countries of the so-called Third World, the private neighbourhoods are marketed as places of a modern and Westernised elite and many inhabitants perceive their place of residence in this way (Caldeira, 2000, p. 263; Glasze, 2003a, p. 162).
- On the supply side, a diffusion of new real estate products can be observed. Coy & Pöhler (2002) and Raposo (2003) report on an 'export' of private and guarded residential complexes from Brazil to Portugal. In Lebanon, several developers of private neighbourhoods knew of this concept from professional experiences outside of the country (Glasze, 2003a, p. 145). And new electronic media like the internet further boost the international spread of new concepts. Similarly to shopping centres, private neighbourhoods are part of a repertoire to which actors of both the demand and the supply side are able to refer.

The examples show that the attractiveness of private neighbourhoods as territorial club economies is increased by several socio-economic and socio-political transformations, often described as effects of globalisation. Nevertheless, the question of whether private neighbourhoods actually do become an individually reasonable option for the actors involved in a specific housing market can only be answered by analysing the historically and geographically differing governance patterns on a national or regional scale.

Formal institutions like laws or contracts as well as informal institutions such as traditional social values and arrangements make urban development path-dependent and increase or lower the attractiveness of private neighbourhoods. This will be illustrated with three short examples: the US, a country of the Third World, the Lebanon and two European welfare states, France and Germany.

The spreading of private neighbourhoods in the US may be interpreted as a continuation and aggravation of 'the culture of privatism' (Judd & Swanstrom, 1997, p. 426), a liberal urban development, which reflects an individualistic concept of democracy. The idea of individual freedom takes the priority over the idea of solidarity (see the discussion between Holzner, 2000 and Priebs, 2000 on the role of public planning in the US and in Germany).

In many countries of the Third World the state does not assure a coherent urban development and is not able to assure basic services. For example, the Lebanese state does not fulfill the idea of an organisational unit, where public authorities have certain autonomy in relation to the particular interests of individual groups and orient their acting at the public interest. Most employees in the public service owe their posts to a confessionally determined preference by a patron. Therefore, they remain clientelistically connected with this patron. The state is used as a tool, which serves to implement particular interests. Public regulations are interpreted as an illegitimate use of the clientelistic structures by participants of another segment and hardly find acceptance as legitimate implementation of a public interest. The confessionally segmented governance patterns lead to a laissez-faire regulation and a weak state supply, creating an environment that makes private neighbourhoods an attractive option for developers and households who can afford to move there (Glasze, 2003d).

The 'non-boom' of private neighbourhoods in the European welfare states may also be explained to a large extent by specific governance patterns. First, the relatively strong position of public planning limits the options of private developments. The developer of a luxurious gated apartment complex on the outskirts of Berlin, for example, failed to establish a private marina for the development as the municipal land use plan prescribed the public access to the lakeside on his private property (Glasze, 2001, p. 40). Second, the idea of 'public space', although a very vague concept, seems to be deeply rooted in the political discourse in the European welfare states. Thus, the developers of a private and gated neighbourhood in the hinterland of the Côte d'Azur had to face a lot of local opposition which has been picked up by regional and national media (Glasze, 2003b, p. 12). Third, as Charmes (2003, p. 107) has shown for the example of France, the regulation of urban development by public authorities seems to find a lot more acceptance than regulations by homeowner associations, and he explains this fact by the dominating idea of a 'republican' political organisation.

Private Neighbourhoods as Shareholder Democracies

In the 1990s, a discussion about the self-administration of private neighbourhoods as a new form of territorial organisation arose in the US. The lobby association of private neighbourhoods, the Community Association Institute, judged the self-administration of private neighbourhoods to be an ideal organisation of local democracy: "... the most representative and responsive form of democracy found in America today".

The decision making in private neighbourhoods follows the model of stock-corporations. In such shareholder democracies, the standards of equality and open decision making are less rigid than in public politics. For stock-corporations this is hardly seen as a problem assuming one dominant joint interest. However, even though most private neighbourhoods are socially quite homogeneous, there are differing interests. There are disagreements between inhabitants and the developer, for example on warranty issues, between absentee owners and inhabitants, for example, on the charges for common facilities, as well as between households with children and households without children, for example, on the construction of a playground. In short: there are politics within private neighbourhoods.

Looking at the constitutions of Western nations, several basic democratic principles organising political life on a national, regional and local scale can be found, for example, the principle of equality, the principle of the sovereignty of the people, as well as the principles of public and pluralistic decision making. In private neighbourhoods, the imbalance between differing interests and missing democratic institutions often leads to the infringement of such basic democratic principles (McKenzie, 1994, p. 122; Scott, 1999, p. 20; Silverman & Barton, 1994, p. 141).

- Violation of the principle of equality: In contrast to public municipalities, the suffrage for the board of directors is not bound to the place of residence and citizen rights but to the property. Therefore, tenants are excluded from decisions concerning their proper neighbourhood. Furthermore, in many countries the voting rights in condominiums and corporations are distributed according to the value of the property: Instead of 'one man one vote', decisions are taken on the basis of 'one Dollar one vote' (Frug, 1999, p. 171).
- Missing 'opposition': There is no institutionalised opposition or any other organisation assuring pluralistic decision making in private neighbourhoods as it is or at least should be assured by parties in a territorial organisation with public municipalities. Thus, the members of the councils have privileged access to information and a privileged power to determine the agenda. Minorities risk being dominated.
- Dictatorial and oligarchic structures: In proprietary neighbourhoods, the former investors keep the majority of the property. Thus they are able to control the development of the open spaces, facilities and services as well as to manage the complex in a profit-oriented way (e.g. to rent out restaurants, shops or leisure-time facilities at maximum costs). However, even in 'normal' homeowner associations the former investors often dominate the decision making by keeping a part of the apartments or houses or by making use of the voting power of friends, relatives or employees.

In view of these deficiencies it is hardly surprising that studies in the USA (Alexander, 1994, p. 148; Blakely & Snyder, 1997, p. 129) and in Lebanon (Glasze, 2003a, p. 238) have found a lot of clashes and frustration within the private neighbourhoods. Consequently, the commitment of the inhabitants to their homeowner association often is very limited.

Are Private Neighbourhoods so Different from Public Municipalities?

Nelson (1989) and Webster (2002, p. 398) have argued that in an economic perspective private neighbourhoods are not as different from public municipalities as it is usually assumed. Regarding the supply side, it has to be said that the differences between municipalities and private neighbourhoods are at least shrinking: municipalities are turning away from direct public provision, instead contracting for these services with private suppliers just as private neighbourhoods do.

Focusing on the consumption, different types of municipalities have to be distinguished with regard to the following two characteristics: first, the financing and second, the socio-economic homogeneity or heterogeneity.

In a pure federal system, where the municipalities are exclusively or predominantly financed by local property or income taxes as in the US, the funding is quite similar to private neighbourhoods: "a set of shared goods is ... financed by a shared cost arrangement" (Webster, 2002, p. 400). Consequently, municipalities with mostly affluent inhabitants are rich and municipalities with mostly poor inhabitants are deprived. The rich municipalities are able to supply collective goods in a higher quantity and quality than the poor municipalities. The wealthy inhabitants who finance these collective goods with their taxes profit from their high quality and quantity not only by consuming but also through stable or increasing home values. Therefore, they are likely to try preventing free-riding by less affluent households, who do not generate 'adequate' tax revenues. If they are able to dominate the decision making of the council, they may use legal instruments to hinder the in-migration of poor households. Danielson has shown that many rich suburban municipalities in the US use exclusionary zoning as such an instrument (1976, p. 1). Their councils establish public regulations that restrict the development of multi-family buildings and thus limit the in-migration of less affluent households. Consequently, the local public goods in these municipalities become quasi club goods. Hence, many rich, small, socially homogeneous and mostly suburban municipalities in the US work de facto much like a private neighbourhood. "Zoning provides the property right [over open spaces, common facilities etc.], local property taxes provide the membership fees, and the city council is de facto a private board of directors" (Nelson, 1989, p. 46). Charmes shows for several examples in France how many of the small and socially quite homogeneous municipalities on the outskirts of metropolitan areas follow a policy that he labels as 'municipal egoism' (2003, p. 134).

With regard to bigger municipalities, Webster is certainly right in stating that "few civic goods are shared equally by all within a city and inclusion and exclusion are facts of urban life" (2002, p. 409). Even a public library or a public swimming pool *has* to exclude in order to avoid over-use and therefore has "club-like consumption characteristics" (Webster, 2002, p. 398). The access may be limited to people living in the specific municipality or regulated by entrance fees.

However, if the financing of the municipalities is not based exclusively or primarily on local sources as it is the case in countries with a centralised system or if there are vertical or horizontal perequations as in the German 'co-operative federalism', the mechanism described above does not work: the quantity and quality of supplied collective goods does not vary enormously from one municipality to another. Therefore, the motivation to prevent free-riding and to exclude less affluent households may still exist but is smaller. In socially heterogeneous municipalities the decision making of the council has to focus not only on the economic interest of the municipality as a whole but also has to balance internally the different interests of the voting inhabitants.

Thus, there seems to be a fundamental difference between club economies established as private neighbourhoods—and some small and homogeneous municipalities as described above—on the one hand and club economies organising the consumption of public facilities in every city on the other hand. In a private neighbourhood, only the people who are able to afford living in the neighbourhood are allowed to use, for example, the recreational facilities. The use of a public swimming pool in a city also has to be regulated, but the rules defining the rights of use are taken in a city council where, ideally, the interests of different groups of the society are represented. Therefore, in socially heterogeneous municipalities with functioning democratic institutions, it is more likely that the city council takes into account the interests of social groups with little economic power. For example, a council might decide that every school class in the city is allowed to use the public swimming pool for 1 hour a week, thus, enabling children with a deprived family background to learn how to swim.

The example of many suburban municipalities in the US shows that it is not appropriate to dichotomise between 'open, democratic and socially balanced' public municipalities and 'closed and secessionist' private neighbourhoods. For that reason, the paper proposes to evaluate case by case to what extent a given territorial organisation assures an efficient supply with collective goods, allows a democratic decision making, provides equal life chances and does not endanger social coherence on a regional or national scale. This summary will try to sketch out such a critical analysis.

The efficiency argument speaks for the establishment of small club economies with a direct connection of shared consumption and collective cost arrangements. Formalised institutions organise the property rights in these club economies and therefore enable a more efficient (private market) supply of local public goods as in heterogeneous and bigger municipalities where there are no such direct and formalised arrangements (Webster, 2002, p. 409). Consequently, several urban economists judge private neighbourhoods as a 'natural evolution' of urban institutions (Foldvary, 1994; Nelson, 1989; Webster, 2002). However, the focus on 'preferences' as the main variable explaining the differences between neighbourhoods blocks from view that "most public services [are] regarded as desirable" (Frug, 1999, p. 171) and that it is the economic and to some extent social and cultural capital which determines the options on the housing market (Whiteman, 1983, p. 346). In the long run, the spread of private neighbourhoods, and of small and homogeneous municipalities, would lead to a territorial organisation where everybody lives in autonomous enclaves according to his financial capacity. The provision with basic services would directly depend on the individual wealth. Basic life opportunities would be distributed in highly unequal ways, even on a local scale. With regard to the question of democracy, there has to be differentiation between the internal decision making and the external politics of private neighbourhoods concerning their social environment. Internally, the social homogeneity, the small scale and the institutionalisation of a neighbourhood organisation may foster a sense of community and voluntary engagement. However, several studies have shown that the internal decision making in private neighbourhoods often violates basic democratic principles. While the basic ideals of contemporary civil democracies are "equal rights and chances for every citizen", the private neighbourhoods can be interpreted as territorial shareholder democracies, which bind political influence and individual life opportunities closely to individual economic strength—the shares in the neighbourhood (Figure 3). Therefore, the political organisation in private neighbourhoods is returning to the days of a census suffrage where political influence was

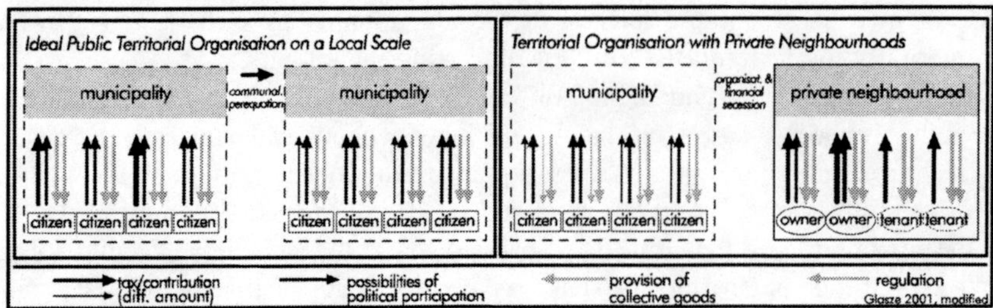

Figure 3. Citizen, public municipality and private neighbourhood: an diagram of territorial organisation. Source: Glasze 2001, modified.

institutionally based on status and class. It must be remembered that Alexis de Tocqueville has labelled the local arena as the 'school of democracy', so the question has to be asked whether the private neighbourhoods are a good place to learn democracy.

With regard to the external politics, the voices seem justified who fear that the politics of socially homogeneous municipalities and private neighbourhoods will be absorbed by the search for an optimal satisfaction of the inhabitants and will not bother with concepts aiming at a social balance on a bigger scale (Frug, 1999; Keating, 1991). The attempts of several homeowner associations in the US and other parts of the world to secede from wider public territorial organisations validate that fear of a 'secession of the successful' (Reich, 1991). The institutionalisation of a new form of a local or sub-local territorial organisation complicates the perequation between wealthy and deprived municipalities and therefore risks (further) complicating the social balance and raising new social barriers.

Conclusion: Club Economies and Shareholder Democracies

The analysis of private neighbourhoods as territorial club economies explains the potential attractiveness of this form of housing for developers and local governments as well as residents. However, the writings relating the club goods theory with the spread of private neighbourhoods tend to overlook the social construction of institutions, the differing interests in society and the unequal distribution of power. Consequently, first, these writings are not able to explain why private neighbourhoods actually do spread in specific regions of the world and not in others and second, they are not able to evaluate the economic, political and social consequences of private neighbourhoods. Therefore, this paper has proposed first to analyse private neighbourhoods as club economies against the background of historically and regionally differentiated patterns of urban governance and second, to critically evaluate the private neighbourhoods as shareholder democracies.

References

Alexander, G. S. (1994) Conditions of 'voice': passivity, disappointment and democracy in homeowner associations, in: S. E. Barton & C. J. Silverman (Eds) *Common Interest Communities: Private Government and the Public Interest*, pp. 145–168 (Berkeley: Institute of Governmental Studies).

Beito, D. T., Gordon, P. & Tabarrok, A. (Eds) (2002) *The Voluntary City. Choice, Community and Civil Society*. (Michigan: The University of Michigan Press).

Blakely, E. J. & Snyder, M. G. (1997) *Fortress America: Gated Communities in the USA* (Washington DC and Cambridge: Brookings Institution Press).

Blandy, S. & Parsons, D. (2003) Gated communities in England, and the rules and rhetoric of urban planning, *Geographica Helvetica*, 58, pp. 314–324.

Borsdorf, A., Bähr, J. & Janoschka, M. (2002) Die Dynamik stadtstrukturellen Wandel in Lateinamerika im Modell der lateinamerikanischen Stadt, *Geographica Helvetica*, 57, pp. 300–310.

Buchanan, J. M. (1965) An economic theory of clubs, *Economica*, 32(125), pp. 1–14.

Caldeira, T. P. R. (2000) *City of Walls: Crime, Segregation, and Citizenship in São Paulo* (Berkeley: University of California Press).

Charmes, E. (2003) *Les tissus périurbains Francais face à la menace des 'gated communties'. Eléments pour un état des lieux* (Paris: Laboratoire CNRS Théorie des mutations urbaines).

Community Association Institute (1999) *National Survey of Community Association Homeowner Satisfaction* (Alexandria: CAI).

Connell, J. (1999) Beyond Manila: walls, malls and private spaces, *Environment and Planning A*, 31, pp. 417–439.

Coy, M. & Pöhler, M. (2002) Gated communities in Latin-American megacities. Case studies in Brazil and Argentina, *Environment and Planning B*, 29, pp. 355–370.

Danielson, M. N. (1976) The politics of exclusionary zoning in suburbia, *Political Science Quarterly*, 91, pp. 1–18.

Feldtkeller, A. (1995) *Die zweckentfremdete Stadt. Wider die Zerstörung des öffentlichen Raums* (Frankfurt/Main and New York: Campus).

Fischer, K. & Parnreiter, C. (2002) Transformation und neue Formen der Segregation in den Städten Lateinamerikas, *Geographica Helvetica*, 57, pp. 245–252.

Foldvary, F. (1994) *Public Goods and Private Communities—The Market Provision of Social Services* (Aldershot, Hants: Edward Elgar Publishing).

Frug, G. E. (1999) *City Making: Building Communities without Building Walls* (Princeton: Princeton University Press).

Ghorra-Gobin, C. (2001) Les espaces publics, capital social, *Geocarrefour*, 76, pp. 5–11.

Giroir, G. (2002) Le phénomène des gated communities à Pékin, ou les nouvelles cités interdites, *Bulletin de l'Association des Géographes Français*, September.

Glasze, G. (2001) Geschlossene Wohnkomplexe (gated communities): 'Enklaven des Wohlbefindens' in der wirtschaftliberalen Stadt, in: H. Roggenthin (Ed.) *Mainzer Kontaktstudium Geographie*, 17, pp. 39–55.

Glasze, G. (2003a) *Die fragmentierte Stadt. Ursachen und Folgen bewachter Wohnkomplexe im Libanon* (Opladen: Leske + Budrich).

Glasze, G. (2003b) L'essor global des complexes résidentiels gardés—atteint ils l'Europe?, *Études Fonçières*, 101, pp. 8–13.

Glasze, G. (2003c) Private neighbourhoods as club economies and shareholder democracies, *BelGeo*, 1, pp. 87–98.

Glasze, G. (2003d) Segmented governance patterns—fragmented urbanism: the development of gated housing estates in Lebanon, *Arab World Geographer*, 6, pp. 79–100.

Glasze, G. & Pütz, R. (2004) Transition, insécurité et marché du logement à Varsovie, *Urbanisme*, 337, pp. 61–63.

Glasze, G., Webster, C. & Frantz, K. (Eds) (2005) *Private Cities— Global and Local Perspectives. Studies in Human Geography* (London: Routledge) (forthcoming).

Gmünder, M., Grillon, N. & Bucher, B. (2000) Gated communities: ein Vergleich Privatisierter Wohnsiedlungen in Kalifornien, *Geographica Helvetica*, 55, pp. 193–203.

Habermas, J. (1990) Vorwort zur Neuauflage, in: J. Habermas (Ed.) *Strukturwandel der Öffentlichkeit*, pp. 11–50 (Frankfurt/M: Suhrkamp).

Holzner, L. (2000) Kommunitäre und 'demokratisierte' Kulturlandschaften: zur Frage der sogenannten 'Amerikanismen' in deutschen Städten, *Erdkunde*, 54, pp. 121–147.

Janoschka, M. (2002) *Wohlstand hinter Mauern: Private Urbanisierungen in Buenos Aires* (Wien: Institut für Stadt- und Regionalforschung).

Judd, D. R. (1995) The rise of the new walled cities, in: H. Liggett & D. C. Perry (Eds) *Spatial Practices: Critical Explorations in Social/Spatial Theory* (London & Delhi: Sage).

Judd, D. R. & Swanstrom, T. (1997) *City Politics: Private Power and Public Policy* (New York: Addison Wesley Publishing Company).

Jürgens, U. & Gnad, M. (2002) Gated communities in South Africa—experiences from Johannesburg, *Environment and Planning B*, 29, pp. 337–354.

Kazig, R., Müller, A. & Wiegandt, K.-Ch. (2003) Öffentlicher Raum in Europa und den USA, *Informationen zur Raumentwicklung*, 3/4, pp. 1–12.

Keating, M. (1991) *Comparative Urban Politics. Power and the City in the United States, Canada, Britain and France* (Aldershot: Edward Elgar).

Landman, K. (2000) An overview of enclosed neighbourhoods in South Africa. *CSIR Report BOU/I* 187 (Pretoria).

Leisch, H. (2002) Gated communities in Indonesia, *Cities*, 19, pp. 341–350.

Lentz, S. & Lindner, P. (2003) Die Privatisierung des öffentlichen Raumes — soziale Segregation und geschlossene Wohnviertel in Moskau, *Geographische Rundschau*, 12, pp. 50–57.

Lichtenberger, E. (1999) Die Privatisierung des öffentlichen Raumes in den USA, in: G. Weber (Ed.) *Raummuster—Planerstoff*, pp. 29–39 (Wien: Eigenverlag des IRUB).

Madoré, F. & Glasze, G. (2003) L'essor des ensemble résidentiels clos en France: un phénomène en expansion et aux ressorts multiples, *Geographica Helvetica*, 58, pp. 225–339.

McKenzie, E. (1994) *Privatopia. Homeowner Associations and the Rise of Residential Private Government* (New Haven and London: Yale University Press).

Mitchell, D. (1995) The end of public space? People's park, definitions of the public and democracy, *Annals of the Association of American Geographers*, 85, pp. 108–133.

Montclos de, M.-A. (1997) *Violence et sécurité urbaines en Afrique du Sud et au Nigeria: Duban, Johannesburg, Kano, Lagos, Port Harcourt* (Paris: L'Harmattan).

Nelson, R.H. (1989) The privatisation of local government: from zoning to RCAs, in: USA C.o.I. Relations (Ed.) *Residential Community Associations: Private Governments in the Intergovernmental System?* A-112, pp. 45–54.

Perouse, J.-F. (2003) La sournoise émergence des cités dites sécurisées en Turquie. Le cas de l'arrondissement de Beykoz (Istanbul), *Geographica Helvetica*, 58, pp. 340–350.

Pöhler, M. (1998) *Zwischen Luxus-Ghettos und Favelas. Stadterweiterungsprozesse und sozialräumliche Segregation in Rio de Janeiro: Das Fallbeispiel Barra da Tijuca.* (Tübingen: Geographisches Institut Universität Tübingen).

Priebs, A. (2000) Raumplanung—Instrument der Obrigkeitsstaatlichkeit oder Instrument einer demokratischen Kulturlandschaft, *Erdkunde*, 54, pp. 135–147.

Raposo, R. (2003) New landscapes: gated housing estates in the Lisbon metropolitan area, *Geographica Helvetica*, 58, pp. 293–301.

Reich, R. B. (1991) Secession of the successful, *The New York Times Magazine*, 20 January, pp. 16–17, 42–45.

Sanchez, T. W., Lang, R. E. & Dhavale, D. (2003) *Security versus Status? A First Look at the Census's Gated Community Data* (Virginia: Metropolitan Institute).

Scott, S. (1999) The homes association: will 'private government' serve the public interest?, in: S. E. Barton & C. J. Silverman (Eds) *Common Interest Communities: Private Governments and the Public Interest*, pp. 19–30 (Berkeley: Institute of Governmental Studies Press).

Siebel, W. (2003) Die überwachte Stadt—Ende des öffentlichen Raums? SWR2 Aula, Manuskriptdienst.

Silverman, C. J. & Barton, S. E. (1994) Shared premises: community and conflict in the common interest development, in: S. E. Barton & C. J. Silverman (Eds) *Common Interest Communities: Private Governments and the Public Interest*, pp. 129–144 (Berkeley: Institute of Governmental Studies Press).

Tiebout, C. M. (1956) A pure theory of local expenditures, *Journal of Political Economy*, 64, pp. 416–424.

Treese, C. J. (1999) *Community Associations Factbook* (Alexandria, VA: Community Associations Institute).

Webster, C. (2002) Property rights and the public realm: gates, green belts and Gemeinschaft, *Environment and Planning B*, 29, pp. 397–412.

Wehrhahn, R. (2003) Gated communities in Madrid: zur Funktion von Mauern im europäischen Kontext, *Geographica Helvetica*, i. pr.

Weiss, M.A. & Watts, J.W. (1989) Community builders and community associations: the role of real estate developers in private residential governance, in: USA C.o.I. Relations (Ed.) *Residential Community Associations: Private Governments in the Intergovernmental System?* A-112, pp. 95–104.

Whiteman, J. (1983) Deconstructing the Tiebout hypothesis, *Environment and Planning D*, 1, pp. 339–353.

Rediscovering the 'Gate' Under Market Transition: From Work-unit Compounds to Commodity Housing Enclaves

FULONG WU
Department of Geography, University of Southampton, Southampton, UK

(Received October 2003; revised June 2004)

KEY WORDS: Gated community, urban China, social inequalities, urban governance, commodity housing

Introduction

Since the discovery of the prototype of the gated residence in North America (Blakely & Snyder, 1997; Low, 2001; McKenzie, 1994), there have been claims of gated communities in virtually every corner of the world (Webster *et al.*, 2002). While the literature emphasises the 'global spread' of gated communities, the physical form of gating has existed for a long time in history. The gate and walls can be dated back at least to the walled city when the city was used for military defence. Traditional Chinese residences have been built in the form of 'courtyard housing', an enclosed living space for extended families. The enlarged version of courtyård housing is the walled city, a typical example is the Forbidden City in Beijing. Walls and gates are essential elements in the structure of the traditional Chinese city, which divided the 'gentry' and the peasants (Skinner, 1977). But it is noted that the socio-spatial division between gentry and 'ordinary residents' (merchants) was characterised by "occupational homogeneity and personal-wealth heterogeneity" (Belsky, 2000, p.59). In other words, within the enclosed form of residences, there was enough diversity, which is different from the relatively homogeneous social areas depicted by the Chicago School in the US. Even in the socialist period when egalitarianism was the predominant ideology, enclosed work-unit compounds were widely built. Gating was not a contentious issue then precisely because it was not a device for social exclusion. However, the meaning of gating has changed or other functions of gating have been rediscovered in the post-reform era.

It would therefore be revealing to ask what is so 'novel' about the fortress residence through studying the Chinese city with a long history of gating. This requires investigation beyond the physical form of gating and probing into the urban experience in different social and economic contexts. What is strikingly similar is the shift towards more market-oriented governance in these different contexts. In Western market economies, gating can be seen as a response to social and economic conditions developed by 'neo-liberalism' (Brenner & Theodore, 2002). This paper uses the context of post-reform 'transition' (Wu, 2003) to examine the changing function of gating so as to shed light on the emergence of the gated community.

The definition of gated community given by Blakely & Snyder (1997) emphasises the restriction of public access to the community. A more comprehensive definition is given by Blandy *et al.* (2003), who define gated communities according to their physical and legal structures as:

> walled or fenced housing developments to which public access is restricted, often guarded using CCTV and/or security personnel, and usually characterised by legal agreements (tenancy or leasehold) which tie the residents to a common code of conduct. (p. 3)

Following this definition, it is not difficult to find examples of gated communities in the world. However, there are two novel aspects in the original US context that deserve attention. First, the fear of crime, which is almost a psychological disorder in a highly fragmented society (Low, 2001) and second, an imposed code of conduct through covenants, conditions and restrictions (CC&Rs) as part of the deed (McKenzie, 1994). These two features make the phenomenon of the gated community sufficiently different from its predecessor of pure defensive space and hint at two major explanations for gated communities: (1) the discourse of 'fear' perpetuated by postmodern urbanism which emphasises diversity; and (2) the club realm of consumption and private governance under the post-Fordist regime of accumulation wherein the Keynesian state retreats from social regulation. This paper will follow these two threads to see to what extent they are applicable in a different context. Particular attention will be paid to the history of community development in urban China. As such, the new commodity housing enclaves are examined beyond the enclosure and in the context of the social geography of the *whole* city. The study highlights that gating is becoming problematic as it becomes an instrument for forging social segregation. The popularity of enclosed enclaves is assessed with reference to post-reform social stratification and the fragmentation of housing consumption.

Club Realm Versus the Discourse of Fear

The concept of 'gated community' embeds a complex tension—as the discourse of 'community' it emphasises shared lifestyles and values which enhance social interaction, yet as a gated space it excludes non-members from social interaction. The literature of the gated community is diverse. Insights have been drawn from a wide range of studies of the conditions under which gated communities have been created. These have been explained through different perspectives: the critique of fortress city (Davis, 1990), the transformation of civil to consumer spaces (Christopherson, 1994), the end of public space (Mitchell, 1995), social polarisation and segregation (Caldeira, 1996), the fear of crime and surveillance (Low, 2001), private governance and homeowners associations (McKenzie, 1994), and the club realm of service delivery (Webster, 2001). Among these explanations, there are two major perspectives: first, seeing the gated community as the club realm between the public and private arena; and second, as a new socio-cultural product of fear and crime avoidance.

Webster (2001) uses institutional economics to explain the formation of shared-consumption goods, and by doing so he sheds light on the 'efficiency' of gating. He argues that the gated community is a club realm of consumption, which is neither public nor private. To its members, the community is shared, just like shopping malls for consumers, but its quality of service is provided at a price, and thus can be delivered with efficiency. The gated community is not seen as the encroachment of the private on the public, in contrast to the 'end-of-public-space' thesis (Mitchell, 1995), but rather is regarded as a 'collective consumption club', or so-called 'proprietary community', bounded by explicitly assigned property rights over neighbourhood public goods. Because of the clear definition of property rights, the risk of degradation of neighbourhood goods is reduced, and "with less risk of congestion by free riders and a heightened sense of co-ownership, neighbourhood goods should increase in quantity and quality", because "households are willing to pay when they move from a traditional neighbourhood to a proprietary neighbourhood", and thus "in this sense they must be seen as an efficient club institution from the point of view of residents" (Webster, 2002, p. 409).

The efficiency of delivery of clubbed goods, however, should not be understood just in pure economic terms within the community itself. More broadly, the 'efficiency' is achieved through excluding not only 'free-riders' but also those who cannot afford to pay. The secession movement in Los Angeles (Keil, 2000) as well as the retreat to the gated community is seen as the revolt of the wealthy. This concurs with the post-Fordist urban condition. The Keynesian welfare state under such a regime is less competent to provide public goods, with the retrenchment of social policies (Peck & Tickell, 1994). On the other hand, the shift toward the economy of scope which emphasises the differences in response to highly differentiated demand means the difficulty of organising standard service provision. Worldwide restructuring towards entrepreneurial governance, wherein services are provided through user charges and public and private partnership, is rooted in the institutional shift which treats citizens as consumers. This creates spaces for the emergence of private governance, of which the gated community is just one type. McKenzie (2003) provides new evidence from Las Vegas, where the municipality compulsorily requires communities to set up their homeowners associations in order to provide public goods by themselves, allowing the municipality to continue to obtain new property tax, thus reducing public expense.

The majority of gated community studies emphasise their origin in the postmodern urban conditions, especially the 'ecology of fear' (Davis, 1990). Gated communities are seen as a response to the fear of crime and the failure of government to ensure adequate security. Worsening security is accompanied by rising social polarisation and residential segregation and consequent social control techniques based on the so-called 'militarisation' of the city (Davis, 1992). Hyper-segregation is commonly found in Latin America where walled settlements have existed for a long time (Caldeira, 1996). Such an architectural form is now popular in the US, forming what Blakely & Snyder (1997) called an 'enclave of fear'. However, Low (2001) points out that the fear of crime is psychological as there is no statistical evidence to show a devastating rise in crime. Thus such a disorder cannot be cured by physical design. Instead, exclusion within the gated community may increase fear of the unknown other, and consequently contributes to residential segregation. The use of fear discourse is to 'naturalise' social and physical exclusionary practice (Low, 2001), and in turn it creates the conditions for generating the fear of otherness. This fear factor has severe impacts on the growth of children.

These two explanations are not mutually exclusive. They may be applied in different degrees to different types of gated community. Blakely & Snyder (1997) classify gated communities into three categories: lifestyle communities, prestige communities and security zone communities. The primary reason for each of these communities is different. Lifestyle communities are more like a club realm of common interests. Prestige communities are more like a product of social differentiation. The security zone community is a response to the deteriorating feeling of security. The function of the gate is different in these communities. In lifestyle communities, the gate serves to define the division of services within and outside the place. In the prestige community, the gate is more symbolic, marking the quality of the environment. This type of community has purely residential use, and thus defining the boundary of services is not an issue. In security zone communities, the gate has the more practical function of enhancing the sense of safety, and often is set up by residents rather than by developers.

In different cultural contexts, however, the primary reasons for gating can be quite different. In a city like São Paulo where social segregation has a long tradition, gated communities seem to have the characteristics of prestige communities (Caldeira, 1996). In Indonesia, for example, the concern for security has always been an issue because of racial tension, but recently exclusion has become primary, stability and privatisation are secondary, and "the sense of community is only tertiary, if any importance at all" (Leisch, 2002, p. 350). The development of gated communities in Jakarta is a response to the need for more exclusive and prestigious lifestyles along with the growth of upper middle classes. Moreover, for different social groups, gating can be developed for different reasons. For expatriates in Beijing, the development of so-called 'foreign housing' resulted from the lack of high-quality commodity housing, which in turn has driven the concentration of expatriates' housing into foreign enclaves (Wu & Webber, 2004). The creation of these gated forms of residence is due to the gap between the demand for such housing and institutional constraints on supply.

Changing Communities in Urban China

Private Housing Areas Built before the Revolution (1949)

Before large-scale industrialisation, Chinese cities shared many features of self-contained rural settlements. The major production unit in the pre-industrial city is the craft workshop mixed with residences. The integration of work and living is common in many pre-industrial Asian cities (Kim *et al.*, 1997). In the early days of industrialisation in the 1920s, the Chinese coastal cities were under Western influences. In Shanghai, Tianjin and Guangzhou, foreign settlements or concession areas were built. Surrounding the factories were early workers' housing, built into terraced houses or 'lane houses' (Lu, 1999). In these industrial cities, the separation between production and residential space began to create different types of urban communities. Social areas based on socio-economic status were constructed. For example, hawkers, small landlords, craftsmen and corner shop owners lived close to the downtown area; industrial workers were accommodated in factory dormitories near factories, and office workers, foreign company employees, petite bourgeoisie and rich local businessmen moved into the 'upper corner' in the foreign concession area. At the same time, the bankruptcy of the rural economy, devastated by civil wars, had driven rural migrants into the cities. In the peri-urban area, and in poor quality areas such as wetland near rivers, scattered shantytowns appeared, as in other Third World cities.

In the urban area built up before 1949, there were no barriers to restrict access to the communities, but in the upper end housing areas, individual villas and houses were enclosed with walls. For example, the Embassy Area in Nanjing was characterised by luxury and enclosed villas. The streets were quiet, indicating a purely residential area. The exclusion occurred at a larger spatial scale as the 'lower classes' lived in other places near downtown. However, the communities were not restricted by walls or gates. The villas were turned into the compounds for senior officials of the Chinese Communist Party when the Peoples' Government of Jiangsu Province and many government departments were established around the area.

Public Housing Areas Managed by the City Government since 1949

The open access of the inner urban area remained after 1949. Unlike the former Soviet Union, China did not see large-scale restructuring after the establishment of state socialism. The tracts of inner-city housing were left untouched except for a few model redevelopment areas. The inner housing area created a special type of urban community, where the housing was managed by the municipal housing bureau in addition to 'residual' private housing with restricted rights to sell (Wang, 1993). The residents mainly worked in small and collectively-owned enterprises which could not provide living quarters for their employees. Because of the lack of investment, redevelopment was rare. Housing demand was mainly met through densification. In these public housing areas, several households had to live in partitioned housing that previously served for a single family, and had to share facilities. Most residents lived there for a long time and neighbourhood interaction was intense, with residents having a strong attachment to the community. Neighbourhood organisations such as the 'residents' committee' instead of the workplace played an important role in governance.

Workplace Living Quarters and Residential Districts

In addition to the inner-city area, two types of residences were developed in the socialist period, with different degrees of access restriction. The living quarters built for a single work-unit, largely associated with industrial development, were developed with clear boundaries or walls that defined the land use. The enlarged version of work-unit compounds is the planned residential district or so-called 'micro-region', a planning concept imported from the former Soviet Union (Grava, 1993). The residential districts are the 'workers' villages' built for several work-units or even for the whole industrial sector. In Shanghai, the 'Chaoyang New Village' is one of the first generation of workers' villages under the workers' housing programme. While these villages generally have open access, sometimes gates have been set up to mark the entrance. Based on the design of the micro-region, the roads inside the community are deliberately enclosed to prevent through traffic. The development of residential districts was encouraged after 1979 when comprehensive development organised by the city began to replace project-based development carried out by individual work-units. Large-scale mixed communities were developed in the peri-urban area. However, because the allocation of housing and service provision were still based on individual workplaces, these mixed residences generally have not developed a strong sense of 'community'.

Along with the changing built form, the system of housing provision has also experienced some changes since 1979 when the economic reform was initiated. The housing provision system in the pre-reform period was characterised by the division between work-unit housing and municipal housing. The policy of housing commodification was introduced in the 1980s (Wang & Murie, 1999). The production of housing was commodified first, whereas the workplaces had been involved in housing provision until 1998. The result was a hybrid approach to housing provision wherein the workplaces purchased commodity housing and then sold houses to staff at a discount price. This practice benefited those who had already obtained favourable housing conditions within work-units. Since 1998, work-units have gradually retreated from in-kind housing allocation. The process of housing commodification has led to differentiation of residential space (Wang & Murie, 2000; Wu, 2002b).

The provision of community services has been commodified. In the socialist period, services were provided by state work-units. The municipality was generally reluctant to invest in the infrastructure as this did not directly contribute to municipal revenue. Community services were defined as a welfare function, but since the 1990s community services have been defined as the tertiary economic sector. Many service functions have been removed from the workplaces and are run by independent companies or the private sector. The discourse of 'big society and small government' reflects such a trend. Work-unit based service provision is criticised for its lack of 'efficiency'. Community services are now redefined as businesses rather than welfare provision. The reduction of welfare provision has been encouraged by the government, as many manufacturing industries suffered from heavy losses. In fact, such a de-industrialisation process is happening rapidly. In Shanghai alone, 1 million workers had been laid-off up to 1999, and registered unemployment reached 174 700 in 1999 (Yin & Lu, 2001).

Commodity Housing Estates

In the late 1990s, 'community building' became the top priority of the government, in response to the demise of work-units and the weakening of the state's 'hierarchical' control. The residents' committee (a mass organisation) and the 'Street Office' (a local government agency) as 'territorial' organisations were consolidated to restore the functionality of governance that had been reduced by population mobility (Wu, 2002a).

In addition, homeowners associations were promoted as experiments in community local democracy. In Shanghai, up to 1999, about 60 per cent of residences had set up homeowners associations (Yin & Lu, 2001, p. 163). In addition to the residents' committees historically based on volunteers, professional property management companies have been introduced into community management to take charge of landscaping, cleaning and security. However, these services are charged for through fees.

The commodification of housing has led to a new type of residence: pure commodity housing estates. These estates are developed by real estate developers and managed by property management companies. To promote an image of high-quality life, the entrances to these estates are often marked by magnificent gates, sometimes in the style of elaborate baroque facades. Some estates adopt so-called 'enclosed property management' which is becoming very popular. Because residents have been filtered through housing affordability, the estate is created as an 'enclave' of those with similar socio-economic status. The housing price is the mechanism of re-sorting residents into different socio-economic status areas. Accordingly, the profile of neighbourhoods is differentiated. At the neighbourhood level, social space is being homogenised, while at the city level different qualities of residence are built, causing residential segregation. The homogenisation and segregation processes have changed the social geography of the Chinese city. This space of differentiation gives a new function to the 'gate' of excluding people of lower socio-economic status. According to the quality of the enclaves, different levels of exclusion are forged.

In terms of the sense of community, the traditional neighbourhood has the strongest attachment. In the work-unit compound, the attachment is based on the affiliation to the workplace and is thus less attached to the place of residence. In the modern residence of workers' villages built for several workplaces, the sense of community is weak as the fragmentation of service provision has exacerbated alienation. In commodity housing enclaves, the residents treat their residence as a space for living rather than a place for social interaction. Not only are these residences relatively new but the residents also have a wide sphere of social activities beyond the community.

Work-unit Compounds

From its appearance, the work-unit compound conforms to the definition of a gated community: the compound is surrounded by walls and gates, secured by the guards employed by the workplace, and the members are affiliated to the workplace and thus have an agreement (although not in legal terms) of acceptable norms. However, the work-unit compound is created for a different reason other than 'self-organised clubs' or security concerns. As a product it is distinct from socialist urbanisation, which emphasises the sphere of production and constrains consumption. The state's growth strategy is forced industrialisation, while urbanisation is regarded as a necessary cost of industrialisation. The provision of housing is thus largely implemented through individual work-units. The so-called socialist public housing system is in fact characterised by many self-contained 'corporate-governed' units.

Why then was such a mode of provision regarded as 'efficient' in the socialist era? First, it is an effective method of overcoming the constraint of inadequate infrastructure. By tying production investment to investment in infrastructure, the state could ensure that the minimum conditions of social reproduction would be met in order to carry out production tasks. Second, the work-unit compound is a quasi-primary society, which saves the cost of information collection and monitoring. Despite its high diversity, all members of the work-unit compound have formed a comprehensive relationship outside the place of residence. Affiliation to the same workplace leads to intensive interaction between residents and thus the management of work-unit compounds is the duty of the estate department and can be easily carried out. On the other hand, from the market perspective, the duplication of facilities that are not open to the outsiders is inefficient. However, by removing intermediate agents such as developers and service providers and their profit motivation, the need for housing can be directly met by supply from the work-unit. This means that, while housing consumption was constrained almost at the standard of necessity, the quasi-primary society allowed the effective satisfaction of housing needs (Webster *et al.*, 2005) without generating homelessness.

The work-unit is thus the basic organisation of 'totalitarian' society, for reasons beyond state dictatorship. In work-unit compounds, because of the integration of different functions in the same place, there is no need for 'surveillance'. It is an area of true 'neighbourhood watch'. Work-unit compounds are more than places for living and have significant implications for social life (Whyte & Parish, 1984). The existence of the work-unit led to the 'organised dependence' suggested by Walder (1986), through which workers became dependent upon organised collective consumption. These compounds are also referred to as 'neo-traditionalism' communities (Walder, 1986) because traditional primary relationships are preserved and replicated in the work-unit compounds.

The need to operate effective resource allocation by overcoming complexity of information gathering, backward infrastructure, and immobile resources produces a 'cellular structure' of collective consumption. This understanding brings us closer to the explanation of the club theory, albeit the cost of the 'transaction' is not produced by market logic. Work-unit compounds serve as a buffer zone between the state and households and are semi-public to affiliated members. Unlike the conception of the city as an anonymous space, the high-level social mix means it is impossible to label the communities as 'working-class', 'public' or 'middle-class' housing areas. Moreover, it is impossible to exclude a particular social stratum (except the peasants, as there was an invisible wall between the city and the rural area through household registration, see Chan, 1994).

Despite the gate and walls surrounding the staff living quarters, the security of work-unit compounds is not stringent. Because of the high social mix, it is difficult to implement identity checks. In most compounds, ordinary urban dwellers are not stopped except for rural migrants who can be judged from their appearance. In a sense, these communities are gated but not fortified. The gate is not closed during the daytime, and the guard serves as an information provider for visitors. The guard also undertakes some minor maintenance and services such as milk and newspaper delivery. For the workers' village that accommodates the employees from several work-units, identity checking is extremely difficult, if not impossible. There are usually several entrances to the residential district; the gate serves more like a decorative built element. Because there are several workplaces, the management is divided between their estate departments. To enhance security, gated compounds are built within the enlarged community. Because security is not conducted by professional property management, the management is more lax. Unlike the commodity housing enclave built by a single developer who in turn requires a property management company to manage the community, 'enclosed property management' is usually not implemented.

In terms of community attachment, work-unit compounds are bounded by all-inclusive relationships formed in the workplace. However, compared with the traditional inner-city housing area, residents are less active in voluntary participation. This is because many activities are sponsored and organised by various departments at the workplace. The work-unit compound is merely an extension of workplace activity space. For compounds containing multiple workplaces, the sense of community is even weaker, because residents may be attached to smaller compounds or buildings and because their workplaces still play a role in the provision of many services.

The study of work-unit compounds suggests that 'gating' cannot be understood only through the physical form. There is a need to understand the social-spatial structure of communities, in particular the composition of residents compared with the composition of those outside the gate, so as to understand the effect of gating.

Commodity Housing Enclaves

The emergence of new enclaves of commodity housing should be understood in the broad context of market transition in China since the late 1970s, in particular housing commodification and differentiation of housing consumption (Lee, 2000; Li, 2000; Wang, 1993; Wang & Murie, 1999; Wu, 1996; Zhou & Logan, 1996). In the suburbs of Chinese cities, a new type of community has emerged in the form of luxury commodity housing estates. These are developed through market-based property development and are often master planned by landscape architects. Commodity housing enclaves present many features of the gated community, such as spatial enclosures with secured gates, walls and fences, security personnel, and contracts with property management companies.

The housing provision system in China has experienced major changes from being one dominated by public housing, especially workplace housing, to one of mixed 'commodity housing', public and private rental housing. At present, in most cities commodity housing projects must obtain land through the land leasing system. The plots of leased land are delineated by the city government, and therefore the construction of commodity housing is allowed to define physical boundaries to separate it from the rest of the city. The differentiation of land price, however, serves as a mechanism for the location of commodity housing estates. Luxury estates are often concentrated into special areas. For example, in Beijing the concentration of affluent residential areas is in the northeast of the city (Hu & Kaplan, 2001) and in Shanghai the high-quality price areas are near the former French Concession, the Hongqiao economic and technological development zones, and Lujiazui and Jiqiao in the new Pudong area. Luxury estates such as Purple Jade, Rome Garden and Riviera in Beijing, and Gubei and Purple Garden in Shanghai, are all exclusively gated.

Commodification of Service Provision

Accompanying the commodification of housing is the commercialisation of property management. In the work-unit compound, the owner of the properties is the workplace, and therefore property management is undertaken by the estate department at nominal or no charge to residents. This subjects property management to bureaucratic control and has led to low maintenance by both the residents and the workplace. In the newly developed commodity housing estate, property management companies take over responsibility for service provision previously undertaken by the local government or state work-units. In fact, these property management companies are often converted from the subsidiaries of housing bureaux in a process of marketisation. The government actively promotes such a change as it helps to reduce administrative costs.

During market-oriented reform, resource decentralisation seriously constrains government-based service provision. Local governments have to find their own resources to support community services. For example, community offices (street offices) have to sign a fiscal contract with the upper level government (the district government). In addition to the administrative budget, which barely covers the basic salaries of local officials, the district government gives their subsidiaries more 'flexible policies', allowing the street offices to operate their own 'sideline' businesses. Against the backdrop of marketisation, street offices have begun to set up enterprises to subsidise administrative expenditure. Through commercialisation of community services, local governments are able to subsidise non-profit social services through various user charges. This practice is called 'using [profit-making] services to subsidise [non-profit making] services', and has paved the way to private governance. The commodification of community services appears in lower status residences as well as in luxury commodity housing estates, but the commodification in the latter is more thorough and fully-fledged.

Property developers use 'packaged' community services to boost the marketing of their estates. Luxury housing estates are often located in suburban areas where municipal facilities are inadequate. The marketing rhetoric not only emphasises a high environmental quality but also stresses comprehensive and high-quality services. The property management company normally takes over such services as rubbish collection, cleaning, greening, security, nurseries, recreation and amenities, even primary, middle and high schools which are traditionally built by the local government. Other social services seen in inner neighbourhoods such as poverty relief and unemployment support are irrelevant to these high-end enclaves, as the high housing prices exclude low-income groups.

Some of the high-level services are naturally exclusive to those who have a high affordability. Therefore, the concentration of customers provides an 'efficient' channel for service delivery. This is more related to the fragmentation of housing markets than to the prevention of 'free-riders': residents in low socio-economic status areas cannot afford the services and thus do not form 'effective' demand. For Riviera in Beijing, a gated community with a high concentration of expatriates, the services are at an exclusively high level: Riviera Country Club offers a wide range of sports and recreational amenities such as a gym, aerobics studio, squash court, tennis courts, indoor and outdoor swimming pools, private massage rooms, Jacuzzis, sauna and steam room, audio and visual theatre, Le Rivierand Café, affiliated membership to a golf centre, and a 24-hour hotline service centre (Wu & Webber, 2004). Other services usually associated with expensive enclaves include beauty salons, nanny and housekeeper agencies, personal financial services and club restaurants. Special services such as international schools are critical to expatriates' enclaves and are usually concentrated in areas of clustered expatriates' villas.

Community Governance

The emergence of commodity housing enclaves has changed the way in which urban communities are managed. The dominant structure of urban governance in the pre-reform period is characterised by state 'hierarchical' control through the work-unit system, and complemented by local governments ranging from municipality, district government, street office, to residents' committees. The street office is not, strictly speaking, a level of government, but rather the representative agency of district government. The residents' committee, however, is defined as a 'voluntary' organisation, namely a 'self-organised mass organisation' according to the organisation rules promulgated by the National People's Congress. However, in reality, the committee undertakes the tasks assigned by the government, such as the maintenance of public order and basic welfare provision. Recently, the residents' committee has experienced 'professionalisation' through which voluntary members are replaced by professional social workers. While there have been strong interactions among residents in inner areas and work-unit compounds, genuine civil society is still weak in urban China (Wu, 2002a).

However, in commodity housing enclaves homeowners associations have begun to compete with residents' committees. According to the regulation of property management in Shanghai, for example, the community is allowed to set up a homeowners association if one of the following three conditions is met: (1) over 50 per cent of the total floor space is sold as commodity housing; (2) over 30 per cent of housing space is sold from public housing; (3) commodity housing has been sold in the community for over two years (Chao and Li, 2001). For many better-off residents, traditional social functions such as social assistance by the residents' committee become irrelevant, and others such as family planning and rural migrant management often cause tension between the two organisations. In the estates of commodity housing, the homeowners association becomes the more powerful organisation, replacing the residents' committee, because the members of the homeowners association are often successful businessmen and the political elite. The association further contracts the management to the property management company. In these communities, governance has experienced the transition from administrative dominance by the residents' committee to self-governance led by the homeowners association. Such a transition occurs in virtually all communities but is more visible in gated commodity housing enclaves.

Social Composition

The social composition of gated commodity enclaves is different from that of work-unit compounds. Table 1 compares the income structure of several gated communities and workers' villages in Shanghai. It can be seen that the income level of community housing enclaves is much higher than that in workers' villages. For the middle-class estates such as

Table 1. Income levels in selected communities in Shanghai (percentages)

Monthly household income (Yuan)	< 800	800–2000	2000–5000	5000–10 000	>10 000
Workers' village					
Quyang village	11	40	39	10	0
Chaoyang village	13	54	33	0	0
Pengpu area	14	. 65	21	0	0
Gated communities					
Haili Garden	0	10	40	50	0
Lujiazui area	4	6	15	35	39
Wanke Garden	5	5	16	32	42

Source: Wang (2001).

Wanke Garden, developed by a prestigious developer originally established in the Shenzhen Special Economic Zone, the income structure is heavily skewed towards the upper end: 42 per cent of households have a monthly income over 10 000 Yuan, and 32 per cent within the range of 5000 to 10 000 Yuan. To put these figures into context, in Shanghai the average annual income per person is 10 988 Yuan. The average household size is 3.08 (SSB, 2000). Therefore, the total average income is about 33 843 Yuan. This means that the income of this group (monthly income over 10 000 Yuan, or annual income over 120 000 Yuan) is at least 3.5 times higher than the average, bearing in mind that the Wanke Garden is still modest among gated communities and far from being a luxurious one.

The concentration of particular social strata can be seen from the case of Nanjing (Table 2). While Mushu Garden, a middle-class estate in the same range as Wanke Garden, still sees a relatively wide spread of government officials, enterprise managers and private-sector entrepreneurs, Dihao ('Empire and Elite Garden') at the luxury end of the market sees exclusive concentration in two sectors: managers of joint-ventures and entrepreneurs. In fact, a 'golden ghetto' such as the Purple Jade Villas in Beijing is purposely designed and targeted at a small fraction of the nouveaux riches (Giroir, 2002). The gated and gentrified enclaves have driven the change in composition of the population, displacing the original residents into the peri-urban area.

Figure 1 shows the changes in the education levels of the original residents and new homeowners in a gated community in Jin'an District of Shanghai. After the setting up of the gated commodity housing estate, the percentage of residents whose level of education was within the range of primary school to junior high school decreased from 50 per cent to 5.8 per cent, while the level of residents with college and university education rose from 17.6 per cent to 58.8 per cent. None of the new property owners are semi-literate or illiterate, while none of the original residents had an education level of or above that of postgraduate. Such a shift in composition clearly reflects the social process accompanying 'gating'. Although gating creates a physical division, the gate itself does not constitute the mechanism for excluding other social groups. Rather, such a shift accompanied by gating should be traced to the commodification of urban development under market transition.

The concentration of particular social groups may be attributed to the deliberate marketing practice of commodity housing as well as affordability-based filtering.

Table 2. Occupation composition of selected communities in Nanjing (percentages, 1998)

Occupation	Nanhu workers' village	Mushu Garden	Dihao (Empire & Elite Garden)
Industrial workers	81.9	0	0
Technical workers	1.6	3.6	0
Government and enterprise officials	3.6	15.3	0
Office workers	10.3	5.7	0
Private & self-employed	0.9	31.0	0
Joint-venture managers	0	5.4	73.6
Private entrepreneurs	0	25.7	25.4
Joint-venture employees	0	13.3	0
Unemployed	1.6	0	0
Others	0	0	1.0

Source: Wu (2001, p. 138).

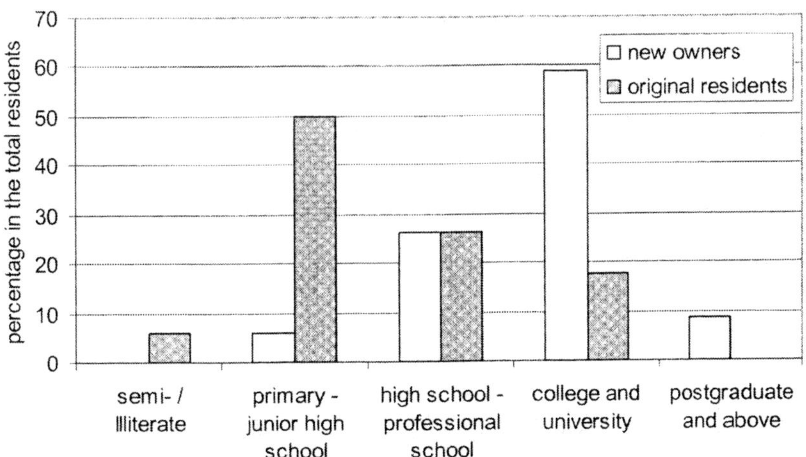

Figure 1. The changing education levels of residents in a gated commodity housing estate in Shanghai, 2002. *Source:* Li (2003).

In Nanjing, the 'Senior Professional Park of Jiangsu Province' is sold to prominent professors and experts according to strict qualification criteria. The villas are sold at a substantial discount price. By engineering a high socio-economic status profile in this enclave, the developer attempts to boost the property price and then sell the properties to the nouveaux riches to regain the profit. In some luxury estates in Beijing, the developer even boasts that, in order to maintain the sense of community, residents are 'screened' before a sale contract can be made, which deliberately excludes some 'private businessmen' without a 'proper occupation' (due to market transition, some business people may make fortunes through illegal activities).

Anonymous Consumer Clubs

The retreat to commodity housing enclaves occurs in the context of the state's failure to provide differentiated services in the public housing area. For expatriates working in multinationals, gated communities provide a familiar living environment which is not available outside (Wu & Webber, 2004). In addition to the security reason, which is obvious for embassy compounds, the formation of expatriate housing enclaves is also attributed to status differentiation. Expatriates earn a much higher income than local employees and have various allowances. 'Foreign housing' is built to a much higher standard than ordinary domestic housing, and the demand for services and facilities differs between the expatriate community and local residents. For many special amenities, such as international schools for expatriates with young children, the clustered gated communities are an efficient form because of the high concentration of customers.

For upwardly mobile residents, the choice of gated community is motivated by the search for the 'good life' in 'privately governed' and anonymous space. The enclosure provides an effective way of organising high-level services and forging status symbols. Very often, these luxury gated communities adopt Western landscape/architectural design, and even have exotic names, such as 'Orange County Beijing'. The transplanted cityscape is in fact using an 'imagined' global motif to open up a venue for commodified housing consumption (Wu, 2004). Exclusive amenities are a particular selling point of these gated communities. Because of economic decentralisation, government's retreat from service provision, and the dismantling of the work-unit system, services have to be purchased. Increasing social stratification, on the other hand, has created differentiated demand for urban living. Gating serves the purpose of effective exploitation of the niche market.

For consumer clubs, the purpose of gating is to demarcate boundaries rather than to build all-inclusive relations. These gated enclaves serve to reduce all-inclusive totalitarian work-unit compounds to anonymous and purified residences. For wealthier residents in gated enclaves, the demise of 'community' should be celebrated because it creates a private space. In Lakeside of Nanjing, double gates erected by the developer not only ensure security but also exclusiveness. Inside the first gate are amenities including a club restaurant, shops and a bowling alley. The guard house is right in the middle, built into a tower overlooking the enclosed place, on the top of which is located the property management company. High security is, therefore, unquestionable. However, some residents interviewed expressed their dislike of entertaining guests in the club restaurant (as business socialising through eating is a tradition in East Asia), although delicious cuisine is served at reasonable prices. The reason is, however, that they are afraid of being spotted.

Anonymity is thus essential to gated communities in urban China. According to a real estate consultant who lives in one such guarded enclave called 'Euro-Classic', this process of achieving anonymity is equivalent to liberation:

> Since I moved into this commodity estate I feel a sort of freedom never felt before. All I know about my neighbour is that he is a film director. He obviously earns a lot of money. He bought two units on this floor and integrated them into one big apartment. Beautiful women, I think probably actresses, are in and out but I don't care. In this area, you don't find old ladies from the residents' committee or curious neighbours keeping an eye on you. (interview, Beijing, 2001)

For the nouveaux riches in the midst of rising social inequality, the retreat into an anonymous life becomes a necessity, so as not to incur jealousy and trouble.

Urban Spatial Structure of China's Gated Cities

Table 3 shows the various reasons for the two types of gated community in urban China. For work-unit compounds, affiliation to the workplace is the primary reason. In fact, such

Table 3. Two types of 'gated communities' in urban China

	Work-unit compounds	Commodity housing enclaves
Affiliation / membership	Primary	None
Housing quality	Tertiary	Primary
Prestige (status)	None	Secondary
Services / property management	Secondary	Tertiary

affiliation can be thought of as a right associated with a job. Housing provision is only one of such bundled rights. Compared with municipal housing, the management of services provided by the workplace is prompt and is usually of better quality. The quality of housing itself is similarly high, as housing is the workplace's asset and empirical evidence shows that workplace housing in general is more favoured than private or municipal housing (Logan *et al.*, 1999). Prestige is irrelevant because of the social mix.

In contrast, the primary reason for relocating into a gated commodity housing enclave is better housing quality assured by higher prices. The search for prestigious and high status areas is always the reason behind the motivation for luxury villa estates. Living in a high-quality environment together with car ownership is now a status symbol. Professional services provided by the management company are also an important consideration. Rather than seeking a communal life, the desire for affiliation or involvement is irrelevant. The need to promote neighbourhood interaction reflects the reality that these communities are purified residences.

The rise of gated commodity housing enclaves has transformed the residential landscape in urban China. The demise of state-organised collective consumption and the emergence of commodified urban spaces create new relations between work and living. Because of weakened ties between workplace and residence, the self-contained workplace compounds are gradually being replaced by fragmented and scattered commodity housing estates. In the residential landscape, there are constellations of luxury housing estates built into gated communities, alongside dilapidated neighbourhoods and migrant enclaves.

Wang & Murie (2000) propose a conceptual model of China's urban structure. Their model consists of three concentric zones: the traditional core, the socialist planned work-unit zone, and a suburban commodity housing estate zone, and focuses on the geography of housing qualities. Different housing estates present a geographical variation in housing qualities but not necessarily in the form of 'concentric' zones. To understand the emerging privately governed urban space and changing social geography of China's gated cities, it is important to understand the different mechanisms of 'spatial sorting'—sorting via state redistribution and sorting through market-based allocation. While the Chinese city consists of a variety of gated residences, their internal social compositions are different. Figure 2 presents the conceptual model of the transition from work-unit compounds to commodity housing enclaves. The three types of housing estates (municipal housing, work-unit housing and commodity housing), although they may not form continuous areas, are differentiated in terms of their social composition.

In the socialist era, the urban structure was composed of the inner city, inherited from the pre-revolutionary era, and the industrial districts built by state-led industrialisation. Private housing dominated the inner city before 1949 and has been transformed into municipal public housing. Some private housing became residual. The municipal housing has generally accommodated those in the reduced informal sector, collectively owned enterprises, and those whose workplaces are too small to provide housing. For those who are affiliated to state work-units, relocation from the previous location to new work-unit compounds or the aggregated version of living quarters (workers' villages) is sorted through the redistributive state. Municipal housing tenants therefore present a lower socio-economic status than work-unit employees, and there is differentiation between different work-units according to their positions (Logan *et al.*, 1999). However, in general, there is a high level of social mix in both inner-city housing and work-unit compounds, as illustrated by the two groups shown in Figure 2.

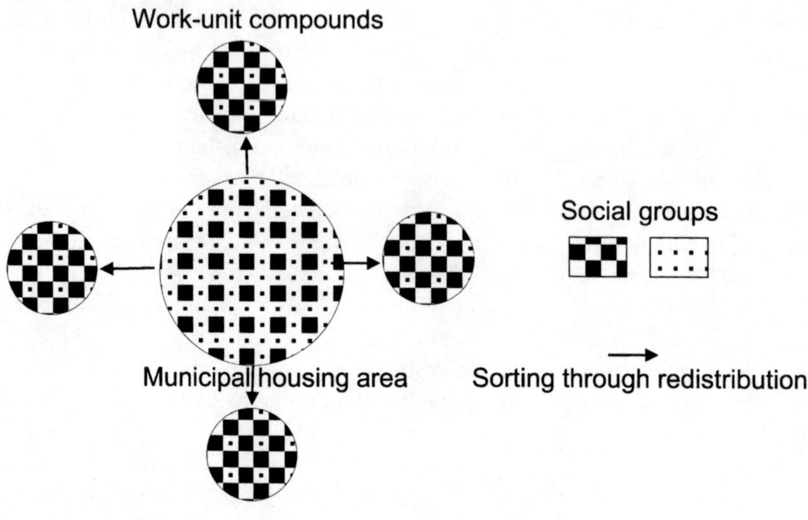

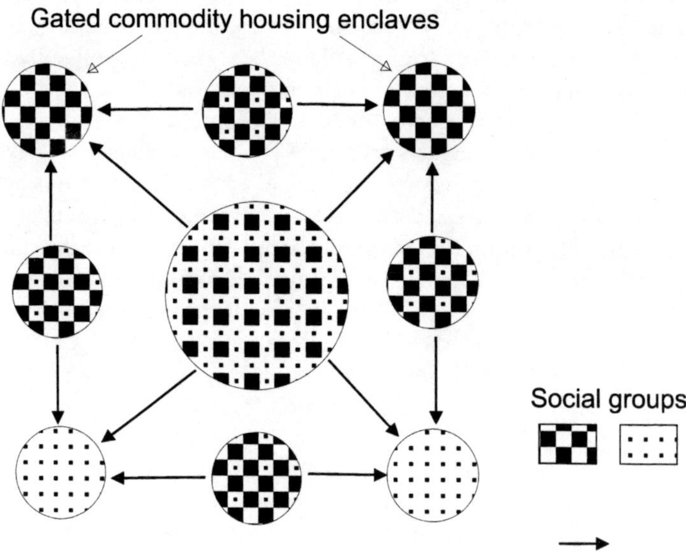

Figure 2. Change in the urban structure characterised by work-unit compounds and commodity housing enclaves.

The gated community adds another component to the above urban structure. The differentiation is now based on the sorting of market-based allocation, i.e. housing affordability. This creates a new pattern of differentiation which is contrasted by the socially-mixed socialist area and segregated post-socialist commodity housing estates. Although the standard of gated communities varies in terms of housing price and the quality of services, the social composition is more homogeneous. The urban structure of commodity housing enclaves becomes more similar to those seen in Western market economies. The concentration of demand further triggers the changing landscape of consumption. Because of the lack of social interaction, the homeowners association has to sponsor various social activities in order to promote residents' participation in these enclaves. However, the effect is not satisfactory: not only because these enclaves are newly built, but also because residents are busy people and their exact motivation is to retreat into the enclave to reduce the 'unnecessary' social interactions of the previous 'totalitarian' society. In these gated communities, the making of 'community' itself is now becoming an impossible mission.

Conclusion: Deconstructing the Space of State-organised Collective Consumption and Reconstructing Post-reform Consumer Spaces

Webster *et al.* (2002), in their editorial entitled 'The global spread of gated communities', highlights the popularity of privately governed residential spaces in different parts of the world, ranging from the Arab world, South Africa, Latin America, to North America and Western Europe. Such a 'spread' should, however, be read beyond the built form of gating itself. Analysis of the political and economic conditions in which gating is created would be illuminating. Strikingly similar is the shift towards more market-oriented urban development in different economies. Gated communities are themselves spaces of new urban governance under neo-liberalism (Brenner & Theodore, 2002). To understand the phenomenon of 'gating', one has to go beyond the gated community itself and examine the social-spatial structure of the city.

While the emergence of 'gated communities' in Southeast Asia has been used as the evidence for the Third Word city's 'convergence to the global city' (Dick & Rimmer, 1998), the built form of 'gating' is not necessarily an imported idea but rather a logical response to changing social conditions. As shown in post-reform urban China, the phenomenon of gating can be dated back to work-unit compounds with a totally different meaning. What is strikingly novel is the selective reshuffling and resorting of households and homogenisation of the community's social composition. In urban China, the function of the 'gate' has been rediscovered, when so-called 'noble communities' are built into commodity housing enclaves.

Needless to say, a gate always exists for the purpose of security. However, what has been revealed in this paper is that gating has been rediscovered with functions beyond the concern for pure security. It draws a line between the dilapidated socialist landscape produced by 'economising urbanisation' and the post-socialist imagination of the 'good life' pursued by upwardly mobile residents. In contrast to the 'fear' factor in the prototype of the fortress-gated community in the US, gating in China's post-reform transition is due more to commodified service provision. Thus, the gate is just a physical manifestation of underlying social fragmentation. While practically the gate may contribute to dividing spaces, it is not the cause of social exclusion.

While the discourse of fear seems less applicable to the Chinese city, urban fragmentation is creating a new urban experience of insecurity in post-reform China. This sense of 'fear' should be understood through the 'deconstruction' and 'reconstruction' of communities in urban China. First, the emergence of modern gated commodity housing enclaves represents a deconstruction process. The work-unit compounds served as an effective medium for state-led industrialisation. The mixture of social composition and self-contained service provision developed a 'totalitarian society' (not in a political sense, but referring to omnipotent social functions). The built form provides opportunities for state ('community') observation and grass roots control. In the process of increasing social inequality and resource marketisation, the nouveaux riches began to feel fear of collectivism and to retreat towards anonymous and 'purified' residences. This is an inversion of the normal fear hypothesis, where in the US the fear of individuals in an 'unorganised' and post-Fordist society leads them to seek a common interest development (CIDs).

Second, 'community building' in the form of gated communities of commodity housing represents a reconstruction process. For those who are detached from state-organised collective consumption, the loss of community surveillance adds a sense of insecurity. Gating is welcomed by both the detached households and the city government. Residents spontaneously seek a solution of secure property management. The state, on the other hand, is helping residents by providing a governance function in 'ungovernable space'. On many occasions, gating is in fact required by the government for building a 'safer living environment'. The Chinese government now actively promotes 'community building', a term more associated with the shift of governance from hierarchical state control to territorial-based management, through the re-organisation of urban space. Commodification of housing and market-oriented resource allocation generates greater mobility and 'higher' urbanism and diversity, as foreseen by Szelenyi (1996) in his hypothesis of 'after-socialist urbanisation', which requires the gated residence.

The novelty of gating in the context where housing consumption has always been organised on a scale smaller than the municipality is that gating changes its social functionality but retains the compelling material reasons for defining the boundary of consumption space. From the club theory point of view, the explanation for the emergence of work-unit compounds is similar to the explanation for commodity housing enclaves. Work-unit compounds reduce the cost of organising collective consumption by the state. Chinese cities have a long history of experimentation with sub-municipal and quasi-public modes of organisation and governance. Work-units contributed to a responsiveness of 'territorial governors' to workers' needs because of strong affiliation and affinity (Webster *et al.*, 2005). In the post-reform era, the gate demarcates emerging consumer clubs in response to the retreat of the state from the provision of public goods. In response to more differentiated affordability, fragmented social status (plus different affiliation relationships with the state, e.g. laid-off workers, migrants, entrepreneurs, expatriates), and diversity of consumption needs, the gated community of commodity housing is seen as an efficient way of organising new consumption space. The conceptual model of the spatial structure of China's gated cities suggests different social geographies in response to changing political and economic conditions, albeit the gate is a common built element. The urban structure is thus characterised by greater fragmentation at the city-wide scale yet also by the homogenisation of microscopic spaces.

Acknowledgements

This research is supported by the project under the British Academy Larger Grant scheme (LRG-37484). I would like to thank anonymous reviewers for their constructive comments. Special thanks are given to Chris Webster who enlightened me during further fieldwork on gated communities in Beijing in March 2004.

References

Belsky, R. (2000) The urban ecology of late imperial Beijing reconsidered: the transformation of social space in China's late imperial capital city, *Journal of Urban History*, 27, pp. 54–74.

Blakely, E. J. & Snyder, M. G. (1997) *Fortress America: Gated Communities in the United States* (Washington DC: Brookings Institution Press).

Blandy, S., Lister, D., Atkinson, R. & Flint, J. (2003) Gated communities: a systematic review of the research evidence. Unpublished report presented at Gated Communities: Building Social Division or Safer Communities? University of Glasgow, 18–19 September.

Brenner, N. & Theodore, N. (2002) *Spaces of Neoliberalism: Urban Restructuring in North America and West Europe* (Oxford: Blackwell Publishers).

Caldeira, T. (1996) Fortified enclaves: the new urban segregation, *Public Culture*, 8, pp. 303–328.

Chan, K. W. (1994) *Cities with Invisible Walls* (Hong Kong: Oxford University Press).

Chao, J. Q. & Li, Z. K. (2001) *Community Management and Property Operation (shequ guanli yu wuye yunzuo)* (Shanghai: Shanghai University Press) (in Chinese).

Christopherson, S. (1994) The fortress city: privatized spaces, consumer citizenship, in: A. Amin (Ed.) *Post-Fordism: A Reader* (Oxford: Blackwell).

Davis, M. (1990) *City of Quartz: Excavating the Future in Los Angeles* (London: Verso).

Davis, M. (1992) Fortress Los Angeles: the militarization of urban space, in: M. Sorkin (Ed.) *Variations on a Theme Park: the New American City and the End of Public Space* (New York: Noonday Press).

Dick, H. W. & Rimmer, P. J. (1998) Beyond the Third World city: the new urban geography of South-east Asia, *Urban Studies*, 35, pp. 2303–2321.

Giroir, G. (2002) The Purple Jade Villas (Beijing): a golden ghetto in red China. Unpublished manuscript.

Grava, S. (1993) The urban heritage of the Soviet Regime: the case of Riga, Latvia, *Journal of the American Planning Association*, 59, pp. 9–30.

Hu, X. H. & Kaplan, D. (2001) The emergence of affluence in Beijing: residential social stratification in China's capital city, *Urban Geography*, 22, pp. 54–77.

Keil, R. (2000) Governance restructuring in Los Angeles and Toronto: amalgamation or secession?, *International Journal of Urban and Regional Research*, 24, pp. 758–781.

Kim, W. B., Douglass, M., Choe, S. C. & Ho, K. C. (1997) *Culture and the City in East Asia* (Oxford: Clarendon Press).

Lee, J. (2000) From welfare housing to home ownership: the dilemma of China's housing reform, *Housing Studies*, 15, pp. 61–76.

Leisch, H. (2002) Gated communities in Indonesia, *Cities*, 19, pp. 341–350.

Li, S. M. (2000) The housing market and tenure decisions in Chinese cities: a multivariate analysis of the case of Guangzhou, *Housing Studies*, 15, pp. 213–236.

Li, Z. G. (2003) Social spatial differentiation in Shanghai. Unpublished fieldwork report.

Logan, J. R., Bian, Y. J. & Bian, F. Q. (1999) Housing inequality in urban China in the 1990s, *International Journal of Urban and Regional Research*, 23, pp. 7–25.

Low, S. (2001) The edge and the centre: gated communities and the discourse of urban fear, *American Anthropologist*, 43, pp. 45–58.

Lu, H. (1999) *Beyond the Neon Lights: Everyday Shanghai in the Early Twentieth Century* (Berkeley, CA: University of California Press).

McKenzie, E. (1994) *Privatopia: Homeowner Associations and the Rise of Residential Private Government* (New Haven: Yale University Press).

McKenzie, E. (2003) Private gated communities in American urban fabric: emerging trends in their production, practices, and regulation. Paper presented at Gated Communities: Building Social Division or Safer Communities?, University of Glasgow, 18–19 September.

Mitchell, D. (1995) The end of public space?, *Annals of the Association of American Geographers*, 85, pp. 108–133.

Peck, J. & Tickell, A. (1994) Searching for new institutional fix: the after-Fordist crisis and the global-local disorder, in: A. Amin (Ed.) *Post-Fordism: A Reader* (Oxford: Blackwell).

Szelenyi, I. (1996) Cities under socialism—and after, in: G. M. Andrusz, M. Harloe & I. Szelenyi (Eds) *Cities after Socialism: Urban and Regional Change and Conflict in Post-socialist Societies* (Oxford: Blackwell).

Shanghai Statistical Bureau (SSB) (2000) *Shanghai Statistical Yearbook 2000* (Beijing: China Statistical Press).

Skinner, G. W. (1977) *The City in Late Imperial China* (Stanford: Stanford University Press).

Walder, A. G. (1986) *Communist Neo-traditionalism: Work and Authority in Chinese Industry* (Berkeley, CA: University of California Press).

Wang, Y. (2001) The study on urban community transition, Unpublished PhD dissertation in Tongji University, Shanghai.

Wang, Y. P. (1993) Private sector housing in urban China since 1949: the case of Xian, *Housing Studies*, 17, pp. 119–137.

Wang, Y. P. & Murie, A. (1999) *Housing Policy and Practice in China* (Basingstoke: Macmillan Press).

Wang, Y. P. & Murie, A. (2000) Social and spatial implications of housing reform in China, *International Journal of Urban and Regional Research*, 24, pp. 397–417.

Webster, C. (2001) Gated cities of tomorrow, *Town Planning Review*, 72, pp. 149–169.

Webster, C. (2002) Property rights and the public realm: gates, green belts, and Gemeinschaft, *Environment and Planning B*, 29, pp. 397–412.

Webster, C., Glasze, G. & Frantz, K. (2002) The global spread of gated communities, *Environment and Planning B*, 29, pp. 315–320.

Webster, C., Wu, F. & Zhao, Y. (2005) China's modern gated cities, in: G. Glasze, C. Webster & K. Frantz (Eds) *Private Neighbourhoods: Global and Local Perspectives* (London: Routledge) (forthcoming).

Whyte, M. K. & Parish, W. L. (1984) *Urban Life in Contemporary China* (Chicago, IL: University of Chicago Press).

Wu, F. (1996) Changes in the structure of public housing provision in urban China, *Urban Studies*, 33, pp. 1601–1627.

Wu, F. (2002a) China's changing urban governance in the transition towards a more market-oriented economy, *Urban Studies*, 39, pp. 1071–1093.

Wu, F. (2002b) Sociospatial differentiation in urban China: evidence from Shanghai's real estate markets, *Environment and Planning A*, 34, pp. 1591–1615.

Wu, F. (2003) Transitional cities, *Environment and Planning A*, 35, pp. 1331–1338.

Wu, F. (2004) Transplanting cityscapes: the use of imagined globalization in housing commodification, *Area*, 36, pp. 227–234.

Wu, F. & Webber, K. (2004) The rise of 'foreign gated communities' in Beijing: between economic globalization and local institutions, *Cities*, 21, pp. 203–213.

Wu, Q. Y. (2001) *Theory and Practice of Residential Differentiation in Large Cities (da chengshi juzhu kongjian fengyi yanjiu de lilun yu shijian)* (Beijing: Science Publisher) (in Chinese).

Yin, J. Z. & Lu, H. L. (2001) *Institutional Reform and Social Transformation (tizhi gaige yu shehui zhuanxing)* (Shanghai: Shanghai Social Science Academy Press) (in Chinese).

Zhou, M. & Logan, J. R. (1996) Market transition and the commodification of housing in urban China, *International Journal of Urban and Regional Research*, 20, pp. 400–421.

Gated Communities in the Metropolitan Area of Buenos Aires, Argentina: A challenge for Town Planning

GUY THUILLIER
Department of Geography, Université Toulouse II-Le Mirail, Toulouse, France

(Received October 2003; revised April 2004)

KEY WORDS: Buenos Aires, gated communities, town planning

Introduction

The massive growth of gated community developments was one of the major urban changes during the 1990s in the Buenos Aires suburbs. In this paper, 'gated community' refers to a closed housing development, to which access is restricted to the residents and their guests. This means that the estate is fenced and gated, and in most cases at least one guard stands at the gate. The biggest gated communities also have patrols controlling the perimeter. If some of them might be collective housing, most of the Argentine gated communities are dedicated to single family housing. In the Metropolitan Area of Buenos Aires today there are about 350 gated communities, covering 300 km^2 of land, and hosting approximately 50 000 permanent residents. Town planners face numerous difficulties because of such developments. Indeed, these upper-class enclaves, requiring huge areas of land, spring up at the fringes of the metropolis, which in Buenos Aires do not consist of 'edge cities' but of slums concentrating the poorest and more recent immigrants in town, coming from the most under-developed provinces of the country. Therefore, a striking contrast appears at the outskirts of the Metropolitan Area, between luxury gated communities and their poorly urbanised vicinity, made up of crowded self-built homes and of pieces of rural land (see Figure 1). Moreover, the municipalities of the second and third rings of the periphery, with high population growth rates and the poorest inhabitants of the whole metropolis, often have limited resources and great difficulties to guide their own urbanisation—when they have the will to do so, which is not always the case. What is the impact of gated communities on poor, fast-growing suburban municipalities? What is the authorities' answer to this phenomenon? What type of juridical and financial tools can local and regional authorities use to integrate gated communities into their urban environment? Which solutions could be explored to reduce the urban and social gaps in the municipalities reached by the spread of gated communities? In an attempt to answer these questions, this paper will first present some facts about the context and extent of this gated urbanisation around Buenos Aires, and its social consequences. Then, focusing on the example of the municipality of Pilar, 50 km north of Buenos Aires, an epicentre for the gated communities' boom, the paper will discuss the impact of these private urbanisations and look at how the local government responds to this new challenge.

The Rise of Gated Communities in Buenos Aires

An Old Story

Gated communities in Buenos Aires have a long history, which begins with a desire for leisure activities and a life closer to nature. During the 19th century, and especially after the terrible epidemics of cholera in 1867 (8000 deaths), and yellow fever in 1871, the wealthy *porteños* (dwellers of Buenos Aires) used to spend the summer in their luxury rural mansions, called *quintas*, away from the crowded and polluted city. Sports clubs with outdoor sports facilities also appeared at the end of the century, as an imitation of the British way of life: during those days, there were many English merchants in Argentina who had key positions in trade and industry. Under their influence, outdoor sports like football, polo, golf, cricket or rowing became increasingly popular amongst the Argentine elite (Troncoso, 2000). At the beginning of the 20th century, railroads, and later cars, contributed to the democratisation of access to a country residence. At that time, chalets and bungalows, that is to say small, cheap country homes, multiplied on the outskirts of the city. They were still often called *quintas*, even though they had little to do with the original great rural domains of the wealthiest population.

The next step was the fusion of sports clubs and country homes, through the formation of the country club, another legacy of the British customs. The first one was the Tortugas Country Club, founded in 1932, 38 km north-west of Buenos Aires, in the territory of the Pilar municipality. Initially, country clubs were leisure-oriented housing compounds, built around excellent sports facilities, such as golf and polo courses, or marinas, and they were inhabited only at weekends and during holidays. During the 1970s, a decade of social unrest and pre-revolutionary troubles in Argentina, important developments and changes took place at the country clubs. Political murders, kidnappings, and brutal military repression created a growing feeling of insecurity and danger that led the upper classes to settle permanently in their country clubs. Most of these gated developments were built

Figure 1. Social structure of the Metropolitan Area of Buenos Aires.

25 to 70 km from the capital, making daily commuting possible. But the real boom of gated communities in the Metropolitan Area of Buenos Aires took place in the second half of the 1990s, for many reasons. For Argentina the 1990s was the time of entering into the global economy. Carlos Menem's government led a vast programme of privatisation of old and inefficient state-run industries and services, attracting important foreign investments. This period of 'easy money' (*plata dulce*) and strong economic growth deepened social inequalities in Argentina, but it was very profitable for the 'winners' (Svampa, 2001) of the new globalised economy, who could then afford to live in a home of their own away from the crowded and busy city centre. The rush for suburbia was encouraged by an intense marketing campaign aimed at persuading the citizens that 'a new way of life', copied from the North-American (sub)urban model, was now possible a few kilometres away from Buenos Aires. For the price of a small flat in a residential tower in the Recoleta, Belgrano or Palermo neighbourhoods (the traditional residence of the upper classes, in the centre of the City of Buenos Aires) now people could enjoy a house in a secure environment, with a small lawn or yard for the children, and maybe, for the wealthiest of them, a golf course, an artificial lake or a polo court nearby.

It must also be emphasised that improvements in the system of suburban motorways were a very important factor in this evolution, the dream of a life closer to 'nature' being now within reach. The constitution of an extensive suburban motorway network was initiated by the military dictatorship (1976–83), who built motorways through the city centre, at the cost of massive urban destruction. In the 1990s, the motorways were extended and enlarged, especially the 'northern access' (*accesso norte*) which leads to the northern suburbs, a favourite place for gated developments.

A Recent Diversification of the Market

As often happens when economic growth is high, the effects of strong speculation were added to these factors, attracting available capital into real-estate investments, and gated developments multiplied on the periphery of Buenos Aires, not only in quantity but also in variety. The old style country clubs spread and democratised through the rise of *barrios privados* ('private neighbourhoods') in the 1990s. *Barrios privados* are inspired by country clubs, but they are cheaper, because their collective amenities and spaces are sharply reduced. *Barrios privados* only offer their residents a house in a fenced and gated estate, sometimes with a tennis court or a small park for children, but without the golf courses or polo courts of the country clubs. Unlike country clubs in their beginnings, *barrios privados* are directly designed for permanent residence. But another housing type also offered a country residence in the 1990s, as country clubs themselves were increasingly used for permanent residence, and were reached by the urban sprawl. Indeed, developers created the *clubes de chacras*, a form of farm-style country homes, usually larger than the ancient country clubs, offering huge plots of 1 hectare or more, whereas plots in country clubs covered 600 to 2000 m^2. For this reason, *clubes de chacras* are beyond the urbanisation front: the 25 *clubes de chacras* of the Metropolitan Area of Buenos Aires are situated at an average distance of 80 km from Buenos Aires, whereas the other gated communities (country clubs and *barrios privados*) are only 45 km on average.

In the 1990s, the boom of gated communities generated so much optimism, profits and speculation that a new type of gated settlement was introduced, on a larger scale: the *megaemprendimientos*. These are huge master-planned communities that include various central and semi-public spaces and amenities (shopping centres, universities, etc.), surrounded by various residential gated neighbourhoods, each one aimed at a specific social and economic profile. Nordelta, the biggest of these projects, in the municipality of Tigre, is 1600 ha wide, and is able to house 100 000 inhabitants. Unfortunately, these grandiose projects were severely struck by the crisis of December 2001, and today the future of *megaemprendimientos* is very uncertain.

A Mass Phenomenon

How can the magnitude of this phenomenon be measured? According to the *Guía de Countries, Barrios Privados y Chacras*, a guide to gated communities published four times since the mid-1990s, in 2000 there were no less than 351 gated communities in the Metropolitan Area of Buenos Aires, covering 300 km^2 of land (see Figure 2). It must be remembered that the surface area of the City of Buenos Aires is only 200 km^2. In fact, one-third of this 300 km^2 of land is dedicated to the huge *clubes de chacras*, but even without the *chacras*, it is a surface equivalent to the City of Buenos Aires that has been fenced.

In 2000, before the great Argentinian crisis, the 351 gated communities were divided into 83 000 plots, with an average size close to 1100 m^2 (without the *chacras*) at an average price of around US$85 000. Only one-third of the existing plots were built, with about 27 000 homes finished, of which only a half (13 000) were inhabited by permanent residents. Even though there are no figures available showing the population of those developments, it is possible, referring to the 13 000 homes built, to estimate that there are 40 000 to 50 000 permanent residents in the gated communities. Standard purchasers of homes in fenced neighbourhoods are young couples with children, therefore it can be estimated that the household size is an average of 3 to 4 persons. Taking into account the owners of country homes, it can be estimated that 80 000 to 100 000 people live in the gated communities in the Metropolitan Area of Buenos Aires.

As stated earlier, the nature of gated communities has changed slightly since they first appeared. Today *barrios privados* are the more common type, with 206 units and only 115 country clubs. Country clubs are the older form, and *barrios privados* were mostly built during the last decade. However, they are usually much smaller than country clubs, since they have no leisure and recreational facilities: if the average size of gated communities in the Metropolitan Area of Buenos Aires is 86 ha, country clubs are 106 ha, whereas *barrios privados* are only 23 ha. However, most permanent residents live in older, larger, and more developed country clubs: the latter still concentrate 75 per cent of the homes and 60 per cent of the permanent population of gated communities in the Metropolitan Area. It must be stressed that the gated communities' population is very concentrated in a few large units: the 30 biggest gated communities in terms of permanent residents have 43 per cent of the total permanent population of the 351 fenced developments. Only 7 gated communities in the 30 largest ones are *barrios privados*, the rest are all country clubs (see Table 1).

It must be noted that this situation is the result of very fast growth, which changed the urban periphery's landscape in just a few years. In 1995, there were only approximately 100 gated communities in the Metropolitan Area of Buenos Aires, covering a surface area of 80 km^2, that is 36 000 plots with 14 000 built houses, of which only 3000 were inhabited by permanent residents. In five years, from 1995 to 2000, the number of gated communities in the Metropolitan Area multiplied by 3.5; the fenced surface area by 3.7;

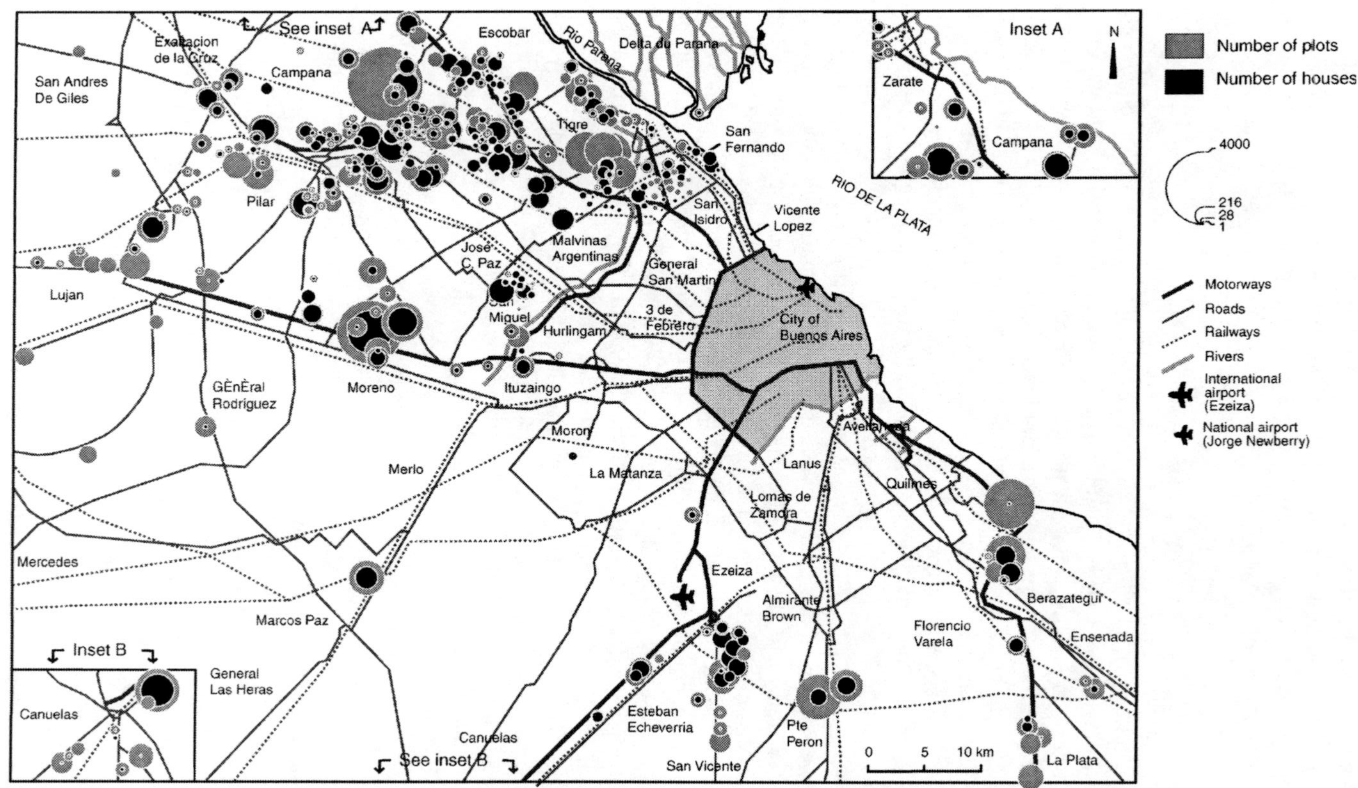

© Guy Thuillier (CIRUS-CIEU), 2002. Source : personal elaboration, based on the datas given by the Guia de countries, barrios privados y chacras, 4 h edition (2000), Publicountry, Buenos Aires.
This map was made with Philcarto, freeware of automatical data cartography developed by Philippe Waniez (download it for free at :http://perso.club-internet.fr/philgeo).

Figure 2. Number of plots and houses built in gated communities in the Metropolitan Area of Buenos Aires (2000).

Table 1. The various types of gated communities (GC's) in the Metropolitan Area of Buenos Aires (year 2000)

Type of gated community	Type of data	Number of developments	Size of development (ha)	Number of plots	Size of plot (m^2)	Price of plot (US$)	Existing houses	% built	Permanent residence	% perm.	Distance to Bs As (km)
Barrios Privados	Average		23	153	928	81 000	34	*23%*	26	*78%*	39
	Total	206	4 691	30 878			6 974		5 433		
	% of all GC's	59%	16%	37%			26%		41%		
Country clubs	Average		106	381	1 281	92 000	171	*47%*	66	*39%*	49
	Total	115	12 132	41 948			19 609		7 638		
	% of all GC's	33%	41%	50%			73%		58%		
Clubes de Chacras	Average		399	168	18 269	90 000	10	*7%*	2	*19%*	85
	Total	25	9 978	3 523			252		48		
	% of all GC's	7%	34%	4%			1%		0,4%		
Megaemprendimientos	Average		733	1 813	1 010	71 000	10	*1%*	3	*30%*	45
	Total	4	2 930	7 250			40		12		
	% of all GC's	1%	10%	9%			0,1%		0,1%		
All GC's together	Average		86	248	2 250	85 000	93	*32%*	38	*49%*	45
	Total	351	28 758	83 749			27 000		13 228		

Notes: The column "% built" indicates the percentage of built plots (built houses/number of lots).
The column "% perm." indicates the percentage of permanent residences (permanent residences/built houses).

Example of reading: There were 206 *barrios privados* in the MABA in the year 2000. 59 % of the gated communities in the MABA are *barrios privados*. The *barrios privados* had an average size of 23 ha, all of them covering a total of 4 691 ha, that is to say 16 % of the whole area occupied by gated communities in the MABA.

Source: Thuillier, 2002, based on information given by the *Guía de Countries, Barrios privados y Chacras*, 4ta edicion, 2000 (Buenos Aires, Publicountry SRL / Federación Argentina de Clubes de Campos).

the number of plots by 2.3; the number of houses by 2, and the permanent population by 4.3 (Thuillier, 2002). This trend was stopped by the severe economic crisis that hit Argentina after December 2001. The gated community market crashed, the sales suddenly stopped. Today, *Megaemprendimientos* seem hugely oversized. Generally, it appears that there has been an over-production of gated developments and many plots will probably never be sold. The Argentinian economy is slowly recovering, but the glorious days of the real-estate boom are gone.

A Fragmented Metropolis?

Inside the Gate: A New Way of Life

The boom in gated communities has important social consequences, changing the distribution of the social groups in the urban space, and creating new forms of contacts and relationships between them. The dominant groups' perception of the city and urban representations and preferences also change. Those who choose to live inside the gates want to experience a new way of life, based on the North-American suburban model. However, it must be borne in mind that for many decades Buenos Aires was a city where the upper class lived in the city centre, in apartment flats, or on the northern outskirts of the city, in the close and dense municipalities of Vicente Lopez, San Isidro and San Fernando (Mora y Araujo, 2000). Leaving the city centre to settle in a gated community in Pilar or Tigre, much further from the city centre, implies a profound change in the urban culture and mentalities. The dense city centre, its concentration of activities, its busy streets and coffee shops, its cultural events and nightlife, are no longer attractive to the successful Argentines. The new suburban residents now value more domestic space, nature and family life, rather than the urban landscapes, the opportunities for recreation and contacts provided by the city. Now they want their own garden, they appreciate being able to 'leave the doors unlocked' when they go out and being able to 'let the children play safely in the streets' (Thuillier, 2001). On the other hand, long commuting distances replace the stresses of living in the city, and make the country residents reluctant to go to the City of Buenos Aires, which slowly becomes stranger and stranger to them. This is particularly true for their children: for some of them, Buenos Aires is an unknown, dangerous and awful place to go. The relationship of those children with the city is now shaped by fear.

Changing places also means changing social interactions. Maybe more valuable than nature, gated communities residents will find new relationships there. They can be sure that their neighbours will belong to the same socio-economic class as themselves. Together with security and a love of nature, the desire for social homogeneity is probably one of the strongest attractions of the gated communities. This dream of a renewed control of the urban landscape and society is also expressed through the rules inside the gated communities, usually very restrictive, and embracing every aspect of urban life. The construction of the houses, public spaces and use of sports facilities, admission of guests, etc., all have rules that aim to guarantee a rigid social order, and are often a source of conflicts between residents, for instance about pets, fencing of private swimming pools, noise or speed limits. This desire for control and transparency goes quite far: in the Mayling Country Club, for example, those who were caught driving over the 30 km/h speed limits, or who had not fenced their swimming pool according to a decision of the neighbourhood council, saw their names written on a board hung in the club-house.

Through these attempts at total control and transparency, gated communities appear as a desperate proposition to fight against the city's chaos and unpredictability, and as a project, finally, to rebuild the meaning of (city) life (Lacarrieu and Thuillier, 2001, 2004).

Outside the Gate: Back to the Wall

For those outside the gates, gated communities have more difficult consequences. As already stated, for a long time the richest *porteños* used to live in the capital city itself, leaving most of the suburbs to poor people, essentially immigrants from the rural regions of the country. Buenos Aires' city centre, for the quality of its urban life and culture, was comparable with the best European cities: Buenos Aires was long reputed to be the 'Paris of Latin America' (Bernand, 1997). However, on leaving the city centre for the suburbs, the landscape changes slowly: housing and public spaces are increasingly run-down from the centre to the periphery. Poor immigrants from the rural and Indian provinces of the north-west, who flocked to Buenos Aires in waves from the 1940s to the 1970s, have settled according to the principles of the *loteo popular*, or 'popular housing': poorly equipped plots on the outskirts of the city, whose owners slowly built their houses themselves or with the help of relatives and neighbours. The development improved little by little, by the lobbying of local associations who fought to get tarmac roads, street lighting, schools and transport and other urban facilities (Clichevsky *et al.*, 1990). But the political and economic troubles that Argentina has been facing since the mid-1970s blocked this system, and the periphery is still far beyond the City of Buenos Aires in terms of urban amenities. The crisis of December 2001 was the final hit: today half the population of the suburbs of Buenos Aires is living below the poverty level, but even before the crisis of 2001, the rate of poverty was around 30 per cent (INDEC).

Thus, the two extremes of the social spectrum in Argentina coexist through the settlement of gated communities amongst this desolate suburban landscape. Opulent gated communities are often surrounded by poor neighbourhoods, or even shanty towns. The social relationships between the two worlds are ambiguous. The gated citizens are aware of the poverty around them: charity clubs, often led by women, raise funds to improve the situation of local poor public schools or hospitals, attended by the lower classes, whereas the residents of gated communities can attend private hospitals, universities and colleges which have followed them from the city centre to suburbia. On the other hand, residents of gated communities are often reluctant to pay their share of local taxes, arguing that they already contract private companies to provide them with the urban services they need, since gated communities are privately owned and run. In Argentina, corruption is so widespread that most of the time people do not believe that their taxes will be used to improve collective amenities and public services.

Even if they often disregard country clubs, the periphery's poorest residents come from all over the metropolis to gated communities, such as Pilar, to try to get a job, men as construction workers or as gardeners, and women as maids or as nurses. Sometimes, some of them, frustrated by their exclusion from all the opportunities of the gated world, try to get their share by force. As it is quite difficult, but not impossible, to burgle inside the gated communities, car attacks have increased on the access roads to some gated developments. The proximity to the motorway is thus made even more valuable for residents of gated communities: the more expensive country clubs are those beside the motorway—a paradox for places supposed to provide a calm and 'rural' shelter away from the city's noise and pollution.

Another paradox is that gated communities create a strange mix of social distances, symbolised by walls and fences, and physical proximity between the richest and the poorest dwellers of the Metropolitan Area of Buenos Aires. The gate itself symbolises this double meaning, for it can be welcoming as well as exclusive. The presence of these islands of wealth amongst a sea of popular neighbourhoods creates frustration and envy, and finally generates the insecurity and violence that gated communities were supposed to remove for their residents. Gated communities are certainly not the cause of the social crisis of Argentina, but they tend to carve in the urban landscape the fractures of a thorn society. How can public authorities integrate those two parts of the dual city and build a better city for all out of this urban mosaic?

Private Neighbourhoods, Public Challenges: The Example of Pilar

Gated Communities' Favourite Place

To deal with the consequences of gated communities on their environment, the example of Pilar is interesting because this municipality is a kind of epicentre for the gated communities' boom in the Metropolitan Area of Buenos Aires (see Figure 3). This city of 233 000 residents in 2001 is situated in the third ring of Buenos Aires' suburbs, about 50 km north-west of the city centre. It must be remembered here that in the spatial distribution of social groups in the Metropolitan Area of Buenos Aires, the more affluent groups live in the City of Buenos Aires and its close surroundings, and the socio-economic level of inhabitants decreases as the distance to the centre increases. The situation is the opposite of what is occurring in North-American cities, where suburbia hosts mainly the upper and middle classes, the inner city being left to the poorest. In Buenos Aires, on the contrary, the further they are from the city centre, the poorer the inhabitants are (Maestrojuan *et al.*, 2000; Torres, 1993). Pilar, with regard to these socio-economic features, appears as a typical city of the third suburban ring. There is no income indicator in the Argentinian census, but other socio-economic features give some information about socio-spatial inequalities inside the Metropolitan Area. In Pilar, families have more children, the population is younger and less educated, housing sanitary conditions are poorer than in the first two rings of suburbs, where the situation is already worse than in the City of Buenos Aires (see Table 2).

However, as well as having disadvantages, Pilar also presents some advantages, at least for the real-estate business: its size, access and situation. First of all, Pilar is a huge municipality by European standards, with a territory of 352 km^2, even if this size is actually quite common for Latin American municipalities. The City of Buenos Aires covers only 200 km^2. Second, Pilar is reached by the 'Panamerican' or 'northern access' motorway, which crosses the municipality at 38 km and then further north at 67 km. This is the perfect distance to the city centre for gated communities: Pilar is far enough from the City of Buenos Aires to offer huge and cheap pieces of land to developers, but close enough to allow daily commuting using the Panamerican motorway. The right distance from the city centre, good motorway access and a lot of available land: these three attractions for gated community development are further maximised by a cultural factor. Indeed, the northern sector of the metropolis has always been the favourite direction for

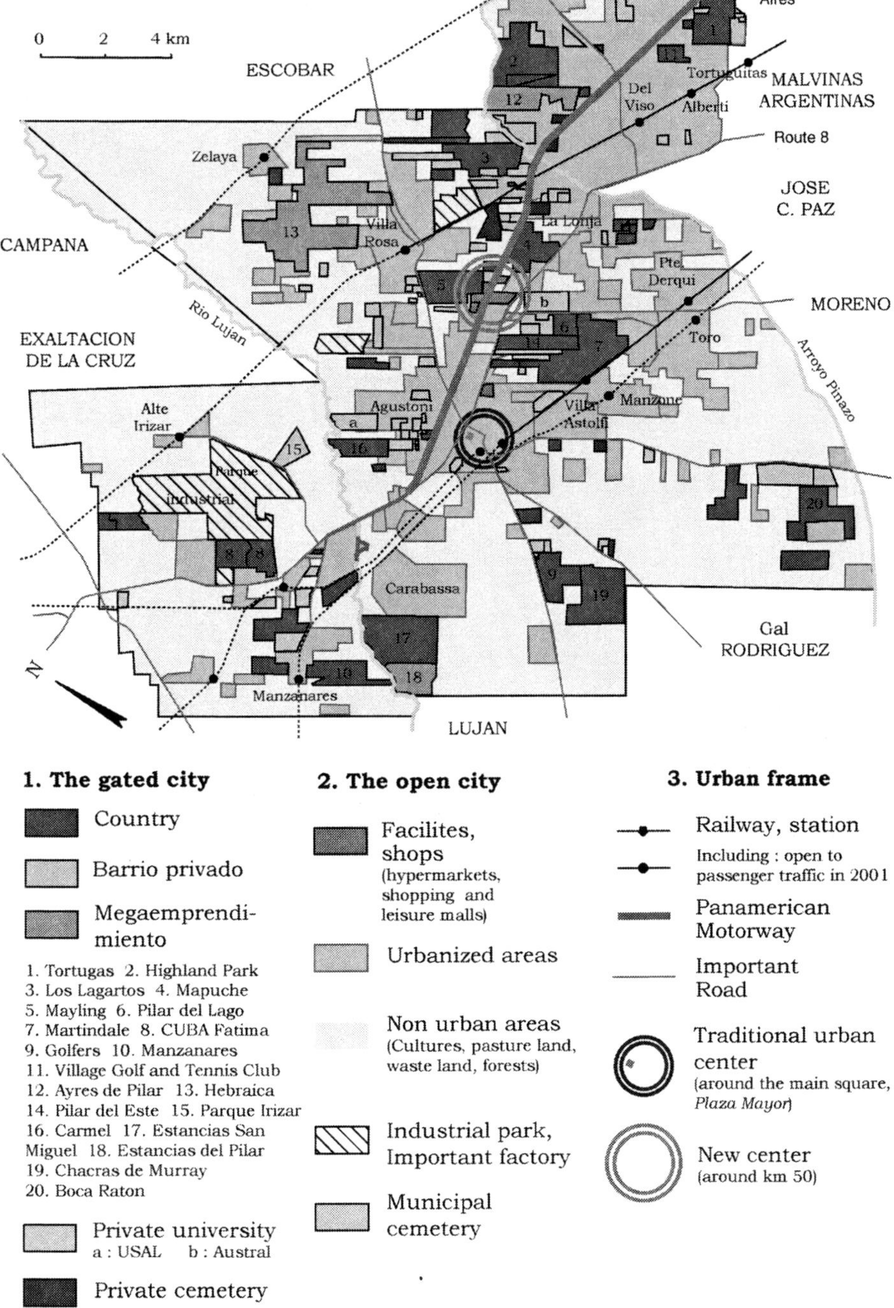

 Personal elaboration, based on datas given by the Municipality of Pilar and the city map of Pilar, AD editions, 2000.

Figure 3. Pilar, a territory under construction.

Table 2. Socio-demographic characteristics of Pilar compared to the rest of Metropolitan Area of Buenos Aires

	Pilar (municipality belonging to 3rd suburban ring)	1st and 2nd suburban rings (24 municipalities)	City of Buenos Aires
Population in 2001	233 000	8 685 000	2 769 000
Population growth 1991–2001	+61%	+9%	−7%
Population under 20 years of age	43%	35%	23%
Fecundity rate (average number of children per woman)	2.2	1.9	1.4
Pop. with only primary education only or no education	72%	69%	41%
Pop. with running drinking water at home (connected to a water distribution network)	18%	58%	98%

Source: Thuillier, 2002, based on INDEC.

the extension of upper-class residence. Throughout history, the affluent *porteños* slowly moved from the city centre to the northern neighbourhoods of the City of Buenos Aires (Recoleta, Palermo, Belgrano etc.) and then to the municipalities of Vicente Lopez, San Isidro and San Fernando, former places of countryside recreation, and then to Tigre and Pilar: the first country club of Argentina, the Tortugas, was created in Pilar in 1932.

All these factors explain why Pilar was at the heart of the gated communities' boom and why today it is the municipality with the highest number of fenced developments in the whole of the Metropolitan Area of Buenos Aires. The city hosts about 120 of them, that is to say about one-third of all gated communities of the Metropolitan Area. These gated neighbourhoods cover 53 km^2, that is to say 14 per cent of the municipal territory. As well as being of a record size, Pilar has another distinction: it is also the city of the Metropolitan Area that had the greatest population growth in the 1990s, increasing from 144 000 residents in 1991 to 233 000 in 2001. This means a demographic growth of 61 per cent, which may be compared with the average growth of the third suburban ring (+21 per cent), of the two first suburban rings (+ 9 per cent) and of all the Metropolitan Area of Buenos Aires (+ 7 per cent) (INDEC).

But in this growth, the population of gated communities itself counts for little: the municipality of Pilar hosts 15 000 residents in fenced developments, that is to say only 6 per cent of its total population. The explanation is that gated communities generated an economic boom for the zone, attracting many qualified and unqualified workers. In the year 2000, there were 8300 houses in gated communities in Pilar. It is thought that the construction of each of them provides 60 jobs for 90 days, and that every inhabited house creates 1.8 permanent jobs (security, maintenance, cleaning, gardening, baby-sitting, etc.) (Marambio, 2000). The municipality of Pilar estimates that gated communities give direct employment to 30 000 persons. Moreover, gated communities have created many indirect jobs, resulting in the economic development they generated in Pilar.

The 'New Pilar'

Trade and services followed the wealthy population to the gated suburbia. Shopping centres and leisure malls are springing up along the northern motorway, for instance at the intersection of km 50 on the motorway in the municipality of Pilar. A few years ago, this place, just a few kilometres away from the city centre, was countryside, except for a handful of large country clubs which were developed there at the end of the 1960s, such as the Mayling Country Club. The area slowly densified, as more country clubs appeared, usually smaller than the pioneers. In 1991, a developer, anticipating the boom in gated communities in Pilar, created a shopping centre at km 50, *Torres del Sol* (the 'Sun Towers'), with 11 000 m^2 of covered surface and 152 expensive shops, whose target was obviously the affluent gated communities' residents. In 1997, in front of *Torres del Sol*, a Village Cine Multiplex with eight cinema theatres, a bingo hall, a bookshop and several restaurants opened. In 1998, on the other side of the motorway another mall was built, with 150 shops, a Jumbo hypermarket of no less than 16 000 m^2, an Easy Homecenter, and a supermarket of home equipment, creating a total of 600 jobs. Nearby, on road 8, there was already a Norte supermarket of 8000 m^2, and moreover, at km 54 on the motorway, The Pilar Design Mall, created by 17 enterprises over an area of 17 000 m^2, offers all the necessary items required for the construction, architecture and design to build a home in a gated community. A new commercial cluster was born in Pilar, competing with the traditional city centre, but clearly oriented to the new residents of the gated communities. Those places, to which access is discretely controlled, are avoided by the poorest residents of Pilar, who do not feel at ease in an environment designed to exclude those who cannot afford high-standard consumption.

However, the retail trade is not the only change brought about in Pilar by the gated communities. At 500 m from km 50, the Farallón group, a developer specialising in gated communities, created Pilar Bureau, an office complex of 12 500 m^2. Two kilometres away, at km 48, a five-star Sheraton Hotel & Convention Centre of 141 rooms was built. In Pilar there was also a huge change in the education on offer: dozens of private schools opened, all offering English and computer science classes. Two private universities have a campus in Pilar: the Universidad del Salvador (USAL), opened in 1987, a campus of 67 ha with an artificial lake, and in 1998 the Austral University created a campus of 70 ha, where in 2000 a university hospital of 45 000 m^2 and 62 beds (later 150) was inaugurated, equipped with the most modern medical facilities. Pilar also has three private cemeteries, including a Jewish one, and they are much better designed and kept in better repair than the public ones.

Challenges for Municipal Action

As stated earlier, gated communities carry the risk of dualisation of the city, acting in opposition to the slums of the rest of the population, but they are also a source of opportunities for municipalities, generating the arrival of a high-income population and important source of potential economic development. How can municipalities control the implementation of gated communities on their territories, what benefits do they get from this type of urban development? Many authors have already pointed out that in Argentina, town planning appears to be a kind of academic exercise with little impact on field reality. There is no national legislation guiding town planning, and the control of the provinces is quite weak. Town planning is roughly left to the municipalities, but the local rules (*ordenanzas*) show many exceptions. Plans and regulations are poorly applied, by a lack of political will and also a lack of concrete means to ensure their applications (Dubois-Maury, 1990).

Once again, gated communities show the validity of these observations. The country clubs have been legally recognised by the Province of Buenos Aires since 1977, under military dictatorship, with the law 8912 of 'territorial organisation and land use'. Before this law, country clubs were built under the status of 'horizontal property', which did not really apply to them, obliging architects and developers to present false plans to the authorities. It can be noted that this same provincial law 8912/77, under the pretext of ecological concern, stopped the 'popular settlements' (*loteo popular*) in the great periphery. The law raised the ecological standards for this form of cheap housing development, thereby making it too expensive and putting it out of reach for the working class. The law 8912/77 also recognised the status of the country clubs, establishing them as recreational compounds, allowing housing for 'transitory residence'. At least 40 per cent of their surface area also had to be dedicated to common sports amenities. The law also indicates a minimum dimension for the plots (600 m^2), a maximum density per hectare (7 to 8 houses), and, amongst other regulations, a minimum distance of 7 km between country clubs. The idea was to avoid the formation of clusters of fenced parcels in the suburban grid. Later, however, the decision concerning this distance was left to municipalities, and in fact, country clubs tended to form themselves together, forming huge areas of closed private land (Verdecchia, 1995).

For *barrios privados*, the basic gated communities without multiple sports amenities, the situation is even worse: there is still no juridical recognition of this form of development. Nevertheless, provincial supervision was improved in 1996 with the creation of a Bureau of Land and Urbanism of the Province, which must give its authorisation to any new development project, on the basis of environmental and social impact studies presented by developers. The municipal authorities must, of course, also give permission for any new development, but real control is weak.

Although there has been a slow improvement in the law, problems often come from its enforcement. Developers complain of the time taken to approve their projects by the provincial Bureau of Land and Urbanism, which can have a two-year delay. For this reason, the construction of a new neighbourhood often begins once the land is purchased, but before obtaining all the necessary permits. It is assumed that it is always possible to find an agreement with the province and the municipality at a later date. In such a context, reported cases of corruption are not surprising. The modernisation of the law and its adaptation to reality today remain the main goal of the powerful FACC (*Federación Argentina de Clubes de Campo*), the Argentine Federation of Country Clubs, which wants to insure its buyers 'juridical security'.

Another problem related to law enforcement is the real capacity of municipal control on urbanisation, which is often questioned. In Pilar, for instance, in 2000 the municipality began a project of 'urban diagnosis' to modernise its land occupation plans which have not been modified since 1985. In those 15 years, the city had nearly doubled its population. Dozens of gated communities were built during this period, under a status of 'exception' to the plans, by special acts of the borough council.

There are also difficult financial problems for the local authorities, as illustrated once again by the case of Pilar. It could be assumed that the municipality with the greatest number of gated communities in the whole of the Metropolitan Area of Buenos Aires would show a comfortable budget surplus, but on the contrary, in 2000 the municipality was close to bankruptcy. Maladministration, waste and corruption might be part of the reasons, but a large number of exemptions were given to shops and industries. Local taxes for gated communities' residents are quite low: they are about eight times lower than the development's service charges. Moreover, tax avoidance is quite high in Argentina, and especially in Pilar: it is thought that only one-third of the due taxes are actually collected. A new municipal authority in Pilar, elected in October 1999, tried to improve the contribution of gated communities to the local budget, but had to reduce its ambitious projects, facing strong pressures from the developers and gated community residents. During the spring of 2000 the *Fundacion Por Pilar* was created, an institution which defines itself as "a group of persons willing to put into action a new policy of integration, development and promotion for all the Pilar district". In fact, this group of people is strongly related to the interests of gated communities, and has prestigious support: it includes VIP members such as Carlos Ruckauf, former governor of the Province of Buenos Aires, and Fernando De La Rua, former president of the Republic, who sometimes plays golf at the Mayling Country Club. Insisting on the connection between gated communities and economic dynamism, the foundation does not hesitate to foresee for the area's future 'Pilar Valley', a kind of Argentinian edge city, a subtle blend of high-tech corporations, gated housing developments and golf greens. But waiting for this grandiose destiny, the real project of this foundation seems rather to put the mayor under tutelage, to make sure that nothing is done against their interests.

Throughout this paper it has been seen that the gated community developments in the Metropolitan Area of Buenos Aires were a massive phenomenon in the last decade, and that they strongly modified the metropolitan landscape. The paper has also shown the risk of moving toward a dual city, where gated communities exacerbate social contrasts, in a period of unbalanced economic growth, thereby creating deepened socio-economic inequalities. This makes a real challenge for town planning and urban governance, for the state and private actors together. Due to a structural lack of real powers and resources of the municipalities, it is clear that controlling this urbanisation causes the greatest difficulties for local authorities, and to follow the dynamism of the private sector, which is boosted by a speculative market. Moreover, the settlement of wealthy people in the area is of little benefit to the poor neighbourhoods surrounding the gated communities, as well as for local governments. On the one hand local political leaders hesitate between the desire to gain more advantages from the fenced developments, but on the other hand, tend to relax the local urban rules and reduce taxes even more, in a context of territorial competition between municipalities to attract the development projects. Developers and residents of gated communities seem to be willing to get more involved in conducting local affairs, as an initiative like the *Fundacion Por Pilar* shows. This certainly comes from a sincere awareness that something must be done to try to smooth the striking contrast between the two sides of the gates in the periphery of the Metropolitan Area. But it also seems to be an attempt to protect their own interests when local politicians start to question the existing order too much. In fact, the efforts of the actors linked to the gated world (developers, administrators, real estate agents etc.) are focused on security problems and on a contained low tax level for gated communities. In the end, building a better city with gated communities probably implies a greater contribution from the developers: for example, some propose a system of compensations, in which a developer, for each fenced development allowed, would have to provide urban facilities for a poor neighbourhood nearby (asphalt on the streets, water and sewerage system, etc.) This proposal certainly also implies stronger and better local, regional and national governments. In a country like Argentina, where corruption is high, and the memory of the dictatorship is still strong, a deepened and more transparent democracy is the first necessary condition for increasing the public authorities' power. Finally, it is understood that a real move towards a more balanced city in Argentina requires a revival of the public government, which in the mean time needs to become stronger and more democratic. Perhaps this goal could be achieved only through a greater mobilisation of the civil society: because Argentina is facing even more dramatic problems, gated communities do not seem to be a big issue in the public debate. However, the crisis of December 2001, and the civic mobilisations and demonstrations that followed, questioning the whole political legitimacy in Argentina, might be the first step onto this long road.

Acknowledgements

This paper summarises some of the findings of a PhD thesis in Geography from the University of Toulouse II (France) in 2002. It is based on a literary review, the analysis of statistical sources and the treatment of their data, but also a field investigation, including observations and interviews with residents of gated and nearby non-gated communities, developers, real estate agents, architects and town planners. The author also wishes to thank Marina Merucci for her revision of the English.

References

Bernand, C. (1997) *Histoire de Buenos Aires* (Paris: Fayard).

Clichevsky, N., Prévôt-Schapira, M. & Schneier-Madanes, G. (1990) *Loteos populares, sector immobiliario y gestión local: el caso de Moreno* (Buenos Aires: CEUR/CREDAL).

Dubois-Maury, J. (1990) Les villes Argentines: une urbanisation sans urbanisme?, *Annales de Géographie*, 556, pp. 695–714.

Guía de Countries, Barrios privados y Chacras (2002) 4th edn (Buenos Aires: Publicountry SRL / Federación Argentina de Clubes de Campos).

INDEC (Instituto Nacional de Estadisticas y Censos) *Censo Nacional 1991, Censo Nacional 2001*. Available at http://www.indec.mecon.ar.

Lacarrieu, M. & Thuillier, G. (2001) Las urbanizaciones privadas en Buenos Aires y su signifación, *Perfiles Latinoamericanos*, 19, pp. 83–113.

Lacarrieu, M. & Thuillier, G. (2004) Une utopie de l'ordre et de la fermeture: 'quartiers privés' et 'countries' à Buenos Aires, *L'espace géographique*, 33, pp. 149–164.

Maestrojuan, P., Marino, M. & De La Mota, G. (2000) *Enclaves urbanos atipicos en el area metropolitana de Buenos Aires* (Buenos Aires: OIKOS).

Marambio, M. (2000) La visión jurídica, in: N. Iglesias (Ed.) *La fragmentación física de nuestras ciudades. Memoria del III séminario Internacional de la Unidad Temática de Desarrollo Urbano, Malvinas Argentinas, 3 y 4 de Agosto de 2000*, pp. 67–70 (Malvinas Argentinas: Municipalidad de Malvinas Argentinas).

Mora y Araujo, M. (2000) Viejas y nuevas elites, in: J. L. Romero & L. A. Romero (Eds) *Buenos Aires. Historia de Cuatro Siglos*, pp. 239–246, 2 vol. (Buenos Aires: Altamira).

Svampa, M. (2001) *Los que ganaron. La vida en los countries y barrios privados* (Buenos Aires: Biblos).

Thuillier, G. (2001) Les quartiers enclos à Buenos Aires: quand la ville devient *country*, *Cahiers des Amériques Latines*, 35, pp. 41–56.

Thuillier, G. (2002) Les quartiers enclos: une mutation de l'urbanité? Le cas de la Région Métropolitaine de Buenos Aires, PhD Thesis in Geography, University of Toulouse II-Le Mirail, France.

Torres, H. A. (1993) *El mapa social de Buenos Aires, (1940–1990)* (Buenos Aires: Facultad de Arquitectura/ Universidad de Buenos Aires).

Troncoso, O. A. (2000), in: J. L. Romero & L. A. Romero (Eds) *Buenos Aires. Historia de Cuatro Siglos*, pp. 93–102, 2 vol. (Buenos Aires: Altamira).

Verdecchia, C. (1995) Los clubes de campo, *Arquis*, 5, pp. 26–28.

Planning Responses to Gated Communities in Canada

JILL GRANT

School of Planning, Dalhousie University, Halifax, Nova Scotia, Canada

(Received October 2003; revised March 2004)

KEY WORDS: Gated communities, planning, Canada

The New Gated Community

Gated or walled communities have proliferated in America in the last decade, and appear increasingly in regions such as the Middle East, Australia, South Africa and Central and South America. Blakely & Snyder (1997) found some 20 000 gated communities in the US accommodating over 3 million units (with 7 to 8 million residents), but more recent estimates (e.g. McGoey, 2003) put the number at more than twice that. The US census of 2001 revealed 7 million households in walled communities, and 4 million households in controlled access communities (Sanchez & Lang, 2002). Developers estimate that 8 out of 10 new residential projects in the US involve gates, walls or guards (Blakely & Snyder, 1997). Some 12 per cent of the population of Metro Phoenix lived in gated communities by 1999 (Webster *et al.*, 2002). Media reports suggest that gated communities are also on the increase in Canada (Anthony, 1997; Haysom, 1996; Liebner, 2003; Yelaja, 2003).

People who choose to close themselves off from the larger city do so in search of community and privacy, and in flight from fear (Dillon, 1994; Hubert & Delsohn, 1996; Low, 2001; Marcuse, 1997; McKenzie, 1994; Wilson-Doenges, 2000). Gates and barriers reflect a reaction to urban problems that have shown no sign of easing; they also indicate the depth of the problems contemporary cities must address. Gated communities respond to the same underlying root issues that generate NIMBYism: concerns about property values, personal safety and neighbourhood amenities (Dear, 1992; Helsley & Strange, 1999; Hornblower, 1988; Rural and Small Town Research, 1992; Shouse & Silverman, 1999). These factors similarly motivate those who consider homes in gated communities. When people feel they cannot rely on public regulations and political processes to protect their neighbourhoods from unwanted uses (or people), then some find the option of voluntarily entering an exclusive community quite desirable (Byers, 2003; Low, 2003).

Gated enclaves represent the hope of security; they appeal to consumers searching for a sense of community and identity; they offer an important niche marketing strategy for developers in a competitive environment; they keep out the unwelcome; they often come associated with attractive amenities; they increase property values (Baron, 1998; Bible & Hsieh, 2001; Blakely, 1999; McGoey, 2003; Townshend, 2002; Webster, 2002). The implications of the growth of this phenomenon are, however, deeply troubling. Gated communities may increase housing costs; they enhance class, age and ethnic segregation; they privatise elements of the public realm (like streets, parks and even schools); they may promote rather than reduce the fear of crime. Are gated communities appropriate in cities seeking to enhance integration and liveability? Gating is clearly profitable, but can it be 'smart' or 'sustainable'?

Blakely & Snyder (1997) indicate that the public debate about the social implications of gated communities has barely begun in the US. The discussion is even more limited in Canada, except perhaps in southern British Columbia (BC) and southern Ontario where most gated projects can be found. However, if Canadian trends parallel those in the US, pressures to gate communities may well increase. Are policy makers and planners ready to respond?

Blakely (2001) says that professional planners have to take an ethical stand on this issue: silence, he argues, implies acceptance of a built realm in which a growing portion of the most affluent among the population wall themselves off. Gated communities raise significant questions related to affordability, segregation and connectivity. They present physical barriers within the community, limiting access to former open landscapes and to public space in coastal areas. As we try to plan sustainable communities with a place for everyone, it might be asked whether gated areas represent an innocuous form of protected suburban development or a worrisome precedent for a divided urban realm.

This paper reports on a study of gated communities in Canada begun in 2002. Relatively little is published on gating in Canada, apart from articles in the popular press. The study sought to develop a greater understanding of the extent of the phenomenon in Canada, the spatial distribution of gated projects, and the character of local planning responses to them. Although the researcher did not think that gated communities were as common as they are in the US, on-going studies of several Canadian cities showed that they were appearing with some frequency. Conducting an inventory of gated projects would provide the researcher with the ability to describe the 'typical' gated enclave in Canada, as well as atypical forms. The sections that follow describe the results of a survey of planners and inventory of gated projects, and discuss planning responses to requests for gating. The sections that follow describe the results of a survey of planners and an inventory of gated projects, and planning responses to requests for gating are discussed. For the most part Canadian planners have paid little attention to gating. Where they have, they often find themselves conflicted about it. Development on private roads (so easily converted to gated enclaves) has become a popular option for municipal governments eager to reduce infrastructure and operating costs in a context where they are expected to provide additional services with few resources (Federation of Canadian Municipalities, 2001).

Gated Communities in Canada

The Canadian planning literature does not include extensive discussions of gated communities. One undergraduate thesis describes a gated project in Winnipeg in limited detail (Golby, 1999), while another recently reported on cases in Nova Scotia (Mittelsteadt, 2003a). Although the Canadian media has highlighted gated projects (e.g. Anthony, 1997; Carey, 1997; Grant, 2003a; Jenish, 2004; Liebner, 2003; Yelaja, 2003), there is little evidence that planners have raised flags about the issue in conferences or professional workshops, prior to recent conference presentations (Grant, 2003b; Maxwell, 2003; Mittelsteadt, 2003b).

Some Canadian scholars have taken an interest in gated projects. Townshend (2002) describes 'self-actualisation' in Canadian retirement communities, some of which are gated; Townshend & Davies (1999) have also explored sense of community in age-related developments in Calgary. Lucas (2002) describes new suburban concentrations of retirement communities; some will be gated. Helsley & Strange (1999) consider the effect of gating on crime, but not particularly in a Canadian context. Byers (2003) explores the way in which gating reveals the fear of the 'other' in society. However, none of these scholars have considered planning issues explicitly: namely, how do local governments seek to control this form of land use?

Although fully gated communities remain relatively uncommon in Canada, the phenomenon is beginning to affect the development industry, especially in the fringe districts of rapidly growing urban areas and in urban and rural regions popular for retirement. In this on-going inventory of Canadian gated communities, over 300 projects have been identified which the evidence suggests have working gates.[1] More than two-thirds of them are in British Columbia. Gated areas have also been located in Alberta, Ontario, Nova Scotia, Saskatchewan and Manitoba. As far as is known, other provinces and territories do not have gated residential projects.

In the fall of 2002, the study began to document gated communities and the planning responses to them. An email survey was sent to planners across Canada, starting with the larger cities and regional capitals, then expanding the search to smaller cities and rural municipalities in growing areas. From 123 contacts, replies were received from 78 planners (63 per cent response rate). In some cases this was followed up with telephone calls for clarification. Since the internet has become a principal marketing tool in the real estate industry today, the internet was also scanned regularly between the fall of 2002 and the end of 2003 for real estate and development listings of any Canadian projects that might have gates, or any homes for sale in gated developments. Web sites typically provide more extensive information than that available in newspaper advertisements, providing important data sought for the study. In key areas, realtors were emailed or telephoned to check on particular communities or to locate additional gated projects.

One of the research team conducted a field study of Nova Scotia projects in fall 2002 (Mittelsteadt, 2003a). In summer 2003, site visits were extended to south central Ontario and to parts of mainland British Columbia. Field work included windshield surveys of residential areas to locate gated developments, visual assessment of some of the gated communities, and interviews with planners, municipal councillors and developers in those key areas. The results herein derive principally from the email survey and field visits.

As soon as the study began it became apparent that planners do not share consensus on the meaning of 'gated'. Although the study focussed on projects with controlled road access, not everyone understood that was implied from this definition: "Gated communities are multi-unit housing developments surrounded by fences, walls or other barriers, and with streets that are not open to general traffic".

Following requests for clarification from early contacts, and submissions of lists of projects that were not considered fully gated, the definition was revised:

> Gated communities are multi-unit housing developments with private roads that are not open to general traffic because they have a gate across the primary access. These developments may be surrounded by fences, walls or other natural barriers that further limit public access.

After a comment from a planner that 'multi-unit' might be interpreted restrictively to mean 'multi-family', a simplified third iteration of the definition was used for the remainder of the study:

> Gated communities are housing developments on private roads that are closed to general traffic by a gate across the primary access. These developments may be surrounded by fences, walls or other natural barriers that further limit public access.

However, despite efforts to make it clear that the study was interested in projects with controlled access roads, planners often used the term 'gated community' to include walled projects with open street access, or to mean developments on private roads. Because of the popularity of walled subdivisions, especially in western Canada, however, it was necessary to be as precise as possible in the terminology used. In completing an inventory the intention was to focus on communities with gates across streets, including those with gates rarely closed.

The study revealed that planners often do not know about gated projects in their midst. Municipal authorities have no system for tracking this development form. In some cases, no permits are required to erect a gate across an entry. Where building permits are required, no one tracks them. Developers or residents' associations can easily add gates to private roads at any time after construction, provided that arrangements are made to give access to emergency vehicles.

Planners often do not know the marketing names of projects, so they had difficulties confirming whether particular developments in their communities were gated. Staff typically track developments by street address, while developers use project names for their marketing. At times, this made it difficult to match up information. Planners living in smaller communities, and those active in local politics, were best able to identify, locate and name gated projects.

When cities were visited for field assessments of gated communities, it was realised that private roads do not appear on street maps. Neither do street maps show marketing names for developments. Accordingly, finding specific projects in the field proved difficult when the researchers did not have a street address from project marketing information. By patrolling likely districts of cities known to have gated projects, and talking to local planners and councillors, the researchers significantly increased the count of gated developments previously identified through the email survey and web search. Due to time and cost constraints, field visits were limited to areas of southern central Ontario, Vancouver suburbs south and west of the city, Vancouver Island, and several communities in the Okanagan Valley. Site visits were conducted in 2002 and 2003.

Web advertising often proved incomplete, and not always accurate. Web sites were searched weekly for gated projects during the summer months of 2003, and at least once per month during fall of 2002, and winter and fall of 2003.[2] Developers may advertise a gate long before it actually goes in, or they may not feature the gate in some of their marketing even though it is there on the ground. Many development web sites are only available during project marketing, and then disappear. Some changed their marketing approaches during the months the researchers were scanning for sites. In some cases, residents' associations established web sites that also provided useful information. Real estate listings were examined online (and where possible also in newspaper ads) for the communities where many gated projects occur to find older projects when re-sale units hit the market. With differing levels and accuracy of information online, it was often difficult to determine the scale or characteristics of gated projects from web information. Efforts to follow up on information with email addresses provided on web sites proved relatively unsuccessful, with low rates of response.

The inventory is partial and incomplete at best: the landscape is changing daily as new projects break ground; gates come and go. Despite the drawbacks, however, the study gives a reasonable approximation of the scope of gated development in Canada. As of early 2004, 314 gated developments had been documented. Based on the rate at which numbers expanded during field studies in BC, the true count may be two to three times that number. Table 1 shows the distribution of known gated projects by province and characteristics.

British Columbia has the greatest number of gated projects, with 228 identified. Three regions have most: Vancouver Island, areas within commuting distance of Vancouver or the Okanagan Valley. Ontario is a distant second with 49 projects. The Ontario projects have greater security, with 9 employing guards and 5 using video surveillance. In total, only 15 have video surveillance, and 15 have guards.

The largest gated community in Canada, Swan Lake in Markham ON, will have 1200 units. It appears that the average size is about 60 units. Few have guards. This makes

Table 1. Documented gated projects in Canada (as of February 2004)

Province	Total gated projects	Projects with 500 units or more	Projects with guards	Population
British Columbia	228	3	5	3 907 738
Alberta	21	3	1	2 974 807
Saskatchewan	8			978 933
Manitoba	1			1 119 583
Ontario	49	8	9	11 410 046
Nova Scotia	7			908 007
Canada total	314	14	15	30 007 094

Note: Other provinces and territories have no confirmed gated projects.

Canadian gated projects much smaller and less security-conscious than their American counterparts. In effect, most are small neighbourhoods; none are full communities. While several include recreational uses, few have other uses (like shops).

There is considerable regional variation in patterns of gating. As far as is known, New Brunswick, Prince Edward Island, Quebec and the north have no gated developments. New Brunswick laws do not allow ground-oriented condominium projects (New Brunswick, 1969), effectively excluding gated forms. Further research is required to clarify why gated projects do not appear in these other regions.

In Nova Scotia and Ontario, gated developments are often located in resort and cottage country, within a two-hour drive of large cities. The Nova Scotia projects are generally relatively small projects with second (or third) homes aimed at a wealthy expatriate market: some advertise their lots and units in American dollars. Some of the Ontario projects are in the suburban fringe, while others locate in more remote areas. Several aim at an adult market. A few prove very expensive and exclusive, with attractive golf courses and other amenities. In general, Ontario has the largest gated developments, with eight having more than 500 units.

Gated projects have become a significant element in the housing market in British Columbia.[3] Gated enclaves are typically in suburban locations, although urban infill sites also occasionally feature gated developments. The data in the current study show considerable variation by community, with gated projects addressing different market segments and resulting in divergent forms and tenures. In high cost areas near Vancouver, adult-oriented townhouse developments proliferate. Where land costs are lower, as in Penticton, 'rancher style' one-storey singles are common. In some areas, mobile home parks offer an affordable gated option. Families are attracted to gated projects in Abbotsford, while seniors fill most projects in Penticton, Langley and Vernon.

Whereas human rights legislation in much of Canada precludes developers from applying age-related restrictions on purchasers, BC and Alberta laws allow age-targeted developments. Because of its moderate climate and beautiful scenery, BC attracts many retirees, making the retirement market a significant component of new development in certain areas. Many projects aim at the seniors or 'adult' market, often including a gate to enhance the appeal. BC communities with a high proportion of elderly residents often have large numbers of gated projects (see Table 2).

Gated projects in Canada invariably involve private roads and condominium (strata or common interest development) or leasehold development. Although some barricading of public streets occurs for traffic control, the study did not find any gated developments with public roads. This situation contrasts with the experience in other countries, where enclosure of public roads may be condoned or permitted.

Table 2. Median age and proportion of seniors in selected cities (2001 census)

Community	Population	No. of gated projects	Median age 2001	% 65+ years 2001
Guelph ON	106 170	1	35.4	12.34
Vernon BC	33 495	45	41.6	20.67
Penticton BC	30 985	10	44.3	25.08
Kelowna BC	96 285	43	40.6	19.16
Canada	30 007 094	308	37.6	12.96

Source: Census of Canada 2001 (Statistics Canada, 2004).

The net densities in gated enclaves may be 20 to 50 per cent higher than in conventional development on public roads: townhouses are at 22 to 26 units per hectare, and singles at 17 to 22 units per hectare (except for high end product). They appear on land zoned 'multi-family residential'. Most of the larger projects feature common amenities such as lavish landscaping, club houses, fountains, RV parking, or swimming pools; some projects, aimed at delivering a lower cost product, have few amenities (other than the gate and wall). Homes within gated developments sell at a premium estimated to be about 10 per cent higher.

The Planning Response

Most planners who responded to the survey said their communities had no gated projects, and no policy to deal with them. The email survey did indicate, however, that 9 of 78 municipalities—five in British Columbia (Burnaby, Coquitlam, Nanaimo, Kelowna and Qualicum Beach) and four in Ontario (Ottawa, Orangeville, Brockville and Ajax)—reported explicit local planning policies to regulate or discourage gated or fortified communities. British Columbia municipalities with several gated projects have the strongest policies in their plans, but face the greatest challenges because of the demand for gating. Field visits indicated that even communities with policies to discourage gating had new gated projects under development.

As Table 3 illustrates, municipalities may use a range of tools to try to manage projects, even when they lack plan policies specifically directed at gating. In the absence of targeted policy on gating, council members and planners may look for other ways to control built form. Policies that limit fence heights, restrict walls or vegetation screens along public roads or require permeable street networks may prevent enclosure. Negotiated development agreements or permits provide planners with mechanisms to discourage developers from gating. In some cases, councils have passed resolutions (rather than adopting plan policies) to ban private roads, limit fortification of properties or prohibit the locking of gates. Planning policies may be among the weakest tools that municipalities have. Other policies, such as requirements for emergency access for fire or police services, may carry greater weight.

Many planners reported that they rely on powers of persuasion to convince developers that proposals for gating are not in the public interest and may therefore slow down applications. As one planner noted:

> There are general policies regarding the preservation of the heritage / culture of the town, which can be raised as a point of discussion with an applicant and identified as a characteristic valued enough by Council to be included in the [plan]. Add to that the persuasion of a good argument, and an applicant can be convinced that the easiest route to achieving smooth planning approval is to concede on certain issues.

Without the local political will to adopt strong plan policies that might prohibit gating, many planners have no choice but to hope that they can persuade developers that gates are not desirable.

Several planners said that their communities experienced little pressure for growth and therefore local developers showed no interest in gated projects. This is certainly the case in most of the country where gated developments are not found. Planners for older inner-city

Table 3. Municipal tools that may be used for controlling gated communities (from email survey of planners)

Tool	Municipality
1. Plan policies and land use / zoning by-laws:	
Adopt explicit plan policies to limit or discourage gating	Burnaby, Coquitlam, Nanaimo, Kelowna, Qualicum Beach BC Ottawa Region (1999)
Restrict use of 'reverse frontage' lots, or require front-loaded lots on all road types	Ajax ON
Limit fence heights	Nanaimo BC
Employ design guidelines (character, heritage, integration of housing)	North Vancouver District
Require or encourage transportation network integration and permeability (may specify grid streets)	Burnaby, Ajax, Orangeville
Require public access	Surrey BC
Set landscaping guidelines or regulations for walls	Regina SK
2. Engineering and emergency access policies:	
Restrict closing of roads, temporary moratorium on private roads	Halifax Regional Municipality NS
Require emergency access	Canmore, Edmonton AB
3. Development agreements and negotiated permitting process adjustments:	
Use development permit process to refuse requests	North Vancouver District, Saanich, Nanaimo Regional District BC
Use urban design and landscape guidelines to limit undesirable features	Toronto, North Vancouver District, Kelowna
Impose deed restrictions or covenants on bare land strata condominiums	Coquitlam
Exact public use easements over private roads or trails	Oakville ON
4. Council by-laws and resolutions:	
Prohibit fortification of buildings and land	Brockville ON
Prohibit locking of gates across roads	Burnaby in 1986
5. Staff persuasion:	
Persuade developers to consider other options	Airdrie AB, Bridgewater NS
Tell developers gates are not permitted	York Region, North Vancouver City
Tell developers staff does not support gating	Cochrane AB, Oakville

municipalities similarly identified little demand for gating. Certainly, the inventory findings do show many of the larger gated communities in the rural and suburban fringe of rapidly-growing urban regions in the west; however, some BC communities, like Vernon, are trying to ensure that gated projects are distributed throughout the city to avoid suburban concentrations. A shortage of developable land in many BC communities limits the scale and location of potential new gated projects to urban infill on 5 to 10 hectare parcels. In these cases, developers eager to create a defined identity for an upscale infill project at a relatively high density want gates. Planners encouraging infill development find themselves torn between achieving density targets and trying to discourage developers from building walls. In BC, the local political climate and culture is such that gated developments with attractive fences have proven an acceptable compromise.

The planners contacted noted several issues that may arise from gated projects. Emergency access for fire, police and ambulance is clearly a concern. Municipalities that have gated projects insist on provisions for emergency vehicles to gain access.

In some municipalities, fire departments work effectively to ensure that roads remain open and wide enough for fire trucks. In other cases, though, emergency personnel have adapted to gates by gaining keys or codes that allow staff to gain access.

Planners generally want to see transportation and pedestrian links maintained wherever possible. Pedestrian and road connectivity has become an important goal. Some planners indicate that they prefer smaller gated projects (under 10 hectares) to large ones. They want to ensure that fences around projects are attractive and allow lines of sight in and out of the project. Landscape and urban design guidelines have become common tools. In rural parts of Nova Scotia, planners note that gating may limit public access to the coastal zone and to areas traditionally used for recreational activities. The survey confirms that municipalities with gated projects are more likely than those without to have developed a policy to deal with potential concerns that follow enclosure.

Some planners worry that gating can lead to social isolation, segregation and fear. In general, they see it as socially undesirable, even though it may be a political reality. As one wrote, "Gated communities are the result of social decay... In summary, they defeat the purpose of community planning". Others, though, can see the appeal of gated projects. One planner explained that her mother lives in a gated seniors community with a good network of friends who look out for her welfare. "Jane Jacobs would be proud", she emailed.

For the most part, planners surveyed did not see a great need to regulate gated communities. Given the press of daily issues, planners have not taken a proactive approach to preventing a phenomenon many have yet to witness in their municipalities. Gated projects have not made it to the 'front burner' in local politics. Where enclaves exist, they are often seen as good neighbours: attractive alternatives for those seeking privacy and quiet. For now, the tools available suit the purpose; others may be added should the need arise.

While most plans have no policies or weak policies to control gating, stronger tools are available where communities wish to limit gated developments. Outright prohibitions were not found. The most effective policies control road networks and walls or fences. Prohibitions on private roads may limit the opportunities of developers and residents' associations to close roads with gates. Plan policies that require street and pedestrian connectivity can establish conditions that restrict road closures. Controls on the height, design, materials, location and extent of walls and fences can affect visual impact, while also preventing outright enclosure. Although these tools are available, for the most part they are not being used. Even where plan policies do explicitly discourage gating, as in parts of BC, municipalities still permit gates.

Perhaps because the culture of fear which drives gating in the US has not proven as strong in Canada, developers perceive a ready market for enclosing new developments only in a few vigorously competitive environments with a high proportion of seniors. Thus planners in many regions have not felt the need to consider the possible implications of and response to gating. With its large share of gated projects, BC has had to consider the question. Planners in some BC communities told us that pressure for gating peaked in the 1980s and 1990s, and has since diminished; in other cities, however, gating proceeds apace. As some planners noted, there is little political appetite to restrict gates when they prove so popular in the market.

The Future of Gating

Canadian provinces show considerable differences in the patterns of gating, as proves true in other countries as well. Regions that attract seniors and wealth tend to attract gates.

Seniors have considerable buying power and clearly form a market interested in gated projects. Many Canadians who head south in winter have experienced gated enclaves in the US, and may appreciate the appeal for privacy, security and prestige. Developers recognise the opportunity for niche marketing: they effectively combine the aesthetic appeal of a private controlled development (with its attractive amenities and common maintenance) with the lure of a homogeneous community of residents: people of similar ages, wealth and lifestyles. The proliferation of 'adult' communities through many parts of southern British Columbia attests to the success of the concept.

Is living in gated communities in Canada primarily about security? The gate is sometimes advertised as a security feature, but the observations here indicate that gates mostly function to keep casual visitors and sightseers out. In some cases, fences are quite low (1.2 m or less). Guards and video surveillance are rare, except in the most exclusive projects. Some gates stand open much of the day. Residents recognise that they are not truly secure, yet those met in the field visits affirm that they watch out for each other and thus reduce random crime. Residents say that they feel safe to travel with some level of confidence as they leave their property tended by neighbours. In Canadian gated projects, privacy, enclosure, identity, lifestyle, homogeneity and community seem most important both to those selling and those buying homes in gated projects. The research does not directly involve interviews with large numbers of residents: understanding their motivations and meanings clearly requires further investigation.

Gating and private roads more generally, offer significant strategies for traffic calming in a context where other options have not worked effectively (Greene & Maxwell, 2004). Gating emergency exit access routes has turned many private roads into veritable cul-de-sacs with no through traffic. Signs typically post lower than normal speeds, and alert motorists that they are entering private property. Parking is carefully controlled. Streets in gated enclaves are narrow, quiet, well maintained, and safe.

Even without operating gates or guards in the guard houses, projects with private roads manage to control access and traffic quite effectively, because most motorists obey signs and stay out. Once private roads are approved, planning rules cannot easily control gating. Private roads can readily be retrofit with gates should greater security at some point prove necessary. In the Canadian context at least, the increasing popularity of development on private roads plays a significant role in the expansion of gated projects.[4]

Where projects are permitted by negotiated agreements, then local government may insert clauses limiting gating. This seems relatively rare in practice. Where gates have not yet become popular, it may be possible to restrict them through the language of development agreements. In some areas, however, they have become so popular among consumers that planners say it is not politically possible to limit them. They have a foothold in the market place.

Canadian municipalities rarely ban gates, although some have passed council resolutions to prohibit locking of gates. More commonly, communities seek simply to regulate the fencing around developments and the size of projects. By controlling fence height, type of materials, articulation and vegetation, local authorities can ensure attractive 'street scapes'. By keeping projects small, they protect connectivity of street and pedestrian routes. Meeting these aims makes gated projects acceptable for many communities.

Planners find themselves torn on gating. On the one hand, many see enclosure as inimical to good planning principles that encourage integration, connectivity, 'eyes on the street', and equality of services. On the other hand, however, they recognise that gated enclaves help meet desired planning objectives for infill development at higher densities. They know that in some cases neighbourhood approval of infill projects may be eased by the promise of enclosure. Affluent developments of adult households are typically seen as good neighbours, unlikely to generate NIMBY responses. Attractive entry features, fences, and landscaping can improve land values in an area. Hence it becomes hard to say 'no' to such proposals.

In part, at least, the proliferation of developments on private roads reflects the funding crisis municipal governments face in Canada. Provincial and federal governments anxious to reduce their own burdens have passed responsibilities on to local governments without providing adequate fiscal resources (FCM, 2001). This financial pinch has made local government vulnerable to cost-saving strategies. Permitting development on private roads saves local government expenditure on road maintenance, snow ploughing, rubbish collection, street lighting, recreational resources and police patrols. With land taxes the major source of municipal revenues in Canada, local governments have a clear incentive to permit development forms that reduce their capital and on-going maintenance costs while providing an enhanced revenue stream. Gated projects of high-end residential developments typically fill with households that place few demands on local governments: households of affluent seniors with their own private recreational resources do not need access to costly schools or other municipal services. Some would suggest that these private developments are 'cash cows' for local governments.

Privatising public services for some may then be justified in terms of devoting the limited resources available to city governments to others with lesser means. Rationalising the trend may let some argue that local governments can concentrate limited resources on providing services to those who cannot participate in private gated clubs (Webster, 2002). However, gating reflects practices that contribute to creating a two-tiered service system and residential segregation in a nation that advocates equality. Recognising the extent of gating in Canada may lead to a public debate about appropriate planning responses to this new urban form.

Acknowledgements

This research is supported by a three-year grant (2002–2005) from the Social Sciences and Humanities Research Council of Canada. An enormous debt of gratitude is owed to research assistants, Lindsey Mittelsteadt, Kirstin Maxwell and Kate Greene for their essential contributions. Jose Canjura, James Bryndza and Craig Walker helped with the inventory. The author would also like to thank the many respondents who volunteered their time to share their knowledge.

Notes

[1] The tally is a best estimate, constantly subject to revision as new information becomes available. Having been first told projects were 'gated', it was later learned from other sources they were not. In some cases, having been told that projects were not gated, a gate was found at the entrance to a development. Gates can be erected or removed on short notice, further complicating efforts to document project status.

[2] Project research assistants working on their thesis research discovered additional projects in early 2004.

[3] BC has the highest proportion of gated developments per capita. In Ontario, there are approximately 0.42 gated projects per 100 000 population; in BC it is estimated there are 5.83 per 100 000. The Canadian ratio is 1.04 per 100 000. Depending on the estimates taken from the US, somewhere between 7 and 14 gated projects are found per 100 000 people there. Thus it can be seen that the BC figures are comparable to the low estimate of American rates of gating, but dramatically higher than those in the rest of Canada.

[4] This research is currently being extended to gain a better understanding of the extent of new development on private roads and the issues related to that phenomenon.

References

Anthony, L. (1997) Safe and sound behind the gate, *Maclean's*, p. 25, 21 July.

Baron, L. M. (1998) The great gate debate, *Builder*, pp. 92–100, March.

Bible, D. S. & Hsieh, C. (2001) Gated communities and residential property values, *The Appraisal Journal*, 69, pp. 140–145.

Blakely, E. J. (1999) The gated community debate, *Urban Land*, 58(6), pp. 50–55.

Blakely, E. J. (2001) Fortifying America: planning for fear, *Planetizen*, Available at www.planetizen.com/oped/item.php?id = 32 (accessed 2 October 2001).

Blakely, E. J. & Snyder, M. G. (1997) *Fortress America: Gated Communities in the United States* (Washington DC: Brookings Institution and the Lincoln Institute of Land Policy).

Byers, M. (2003) Waiting at the gate: the new postmodern promised lands, in: H. Bartling & M. Lindstrom (Eds) *Suburban Sprawl: Culture, Ecology and Politics* (Lanham, MD: Rowman and Littlefield).

Carey, E. (1997) Metro joins trend to guarded communities, *The Toronto Star*, p. A1, 15 June.

Dear, M. (1992) Understanding and overcoming the NIMBY syndrome, *Journal of the American Planning Association*, 58, pp. 288–300.

Dillon, D. (1994) Fortress America, *Planning*, 60, pp. 8–12.

Federation of Canadian Municipalities (FCM) (2001). Municipal finance. (A position paper adopted at the Annual Conference.) Available at www.fcm.ca/english/national/finance1.htm (accessed 26 February 2004).

Golby, J. (1999) Gated communities: the fortress frontier, BA Thesis, Department of Geography, University of Winnipeg.

Grant, J. (2003a) Is there a gate in your future?, *The Chronicle Herald/Mail Star*, p. B2, 14 August.

Grant, J. (2003b) Canadian planning approaches to gated communities. Paper presented at the Canadian Institute of Planners Conference, Halifax, 6–9 July.

Greene, K. & Maxwell, D. K. (2004) Taking matters into their own hands: traffic control in Canadian gated communities, *Plan Canada*, 44(2), pp. 45–47 June.

Haysom, I. (1996) Gated communities on the increase—but no armed guards yet. (Southam newspapers.) *Background in depth, National and international news*, 14 February. Available at http://www.southam.com/nmc/waves/depth/homes/home0214a.html.

Helsley, R. & Strange, W. (1999) Gated communities and the economic geography of crime, *Journal of Urban Economics*, 46, pp. 80–105.

Hornblower, M. (1988) Not in my backyard, you don't, *TIME*, pp. 58–59, 27 June.

Hubert, C. & Delsohn, G. (1996) In age of unease, some appreciate Big Brotherly vigilance, *Sacramento Bee*, 3 December, http://www.sacbee.com/news/projects/terror/bigbrother.html.

Jenish, D. (2004) Can fences make good neighbors?, *Forum*, 28, pp. 21–23 (Canadian Federation of Municipalities).

Liebner, J. (2003) Downsizers find community spirit at London site, *Toronto Star online*, 4 January 2003. Available at http://torontostar.com/NASApp/cs/ Content(d = 1035776145117 and call_pageid = 968332188492 (accessed 18 January 2003).

Low, S. M. (2001) The edge and the center: gated communities and the discourse of urban fear, *American Anthropologist*, 103, pp. 45–58.

Low, S. M. (2003) *Behind the Gates: Life, Security, and the Pursuit of Happiness in Fortress America* (New York: Routledge).

Lucas, S. (2002) From Levittown to Luther Village: retirement communities and the changing suburban dream, *Canadian Journal of Urban Research*, 11, pp. 323–342.

Marcuse, P. (1997) Walls of fear and walls of support, in: N. Ellin (Ed.) *Architecture of Fear*, pp. 101–114 (New York: Princeton Architectural Press).

Maxwell, K. (2003) The legal framework for addressing gated communities in Canada. Paper presented at Canadian Institute of Planners Conference, Halifax, 6–9 July.

McGoey, C. E. (2003) Gated community: access control issues, *Crime Doctor*. Available at http://www.crimedoctor.com/gated.htm (accessed 24 May 2003).

McKenzie, E. (1994) *Privatopia: Homeowner Associations and the Rise of Residential Private Government* (New Haven: Yale University Press).

Mittelsteadt, L. (2003a) Gated communities in Nova Scotia, Thesis, Master of Urban and Rural Planning, Dalhousie University.

Mittelsteadt, L. (2003b) A case study of gated communities in Nova Scotia. Paper presented at Canadian Institute of Planners Conference, Halifax, 6–9 July.

New Brunswick, Province of (1969) *Condominium Property Act*, chapter C-16.

Ottawa, Region of (1999) *Ottawa Regional Official Plan*. Ottawa.

Rural and Small Town Research (1992) *Guidelines for Action: Understanding Housing-Related NIMBY* (Amherst NS: Rural and Small Town Studies Program, Mount Allison University).

Sanchez, T. W. & Lang, R. E. (2002). Security versus status: the two worlds of gated communities. Draft Census Note 02:02, November, Metropolitan Institute at Virginia Tech.

Shouse, N. & Silverman, R. (1999) Public facilities in gated communities, *Urban Land*, 58, p. 54.

Statistics Canada (Statcan) (2004) Age and Sex, 2001 counts for both sexes, for Canada and census subdivisions. 2001 Census Online www12.statcan.ca/english/census01/products/highlight/AgeSex (accessed 27 February 2004).

Townshend, I. (2002) Age-segregated and gated retirement communities in the third age: the differential contribution of place-community to self-actualisation, *Environment and Planning B: Planning and Design*, 29, pp. 371–396.

Townshend, I. J. & Davies, W. K. D. (1999) Identifying the elements of community character: a case study of community dimensionality in old age residential areas, *Research in Community Sociology*, 9, pp. 219–251.

Webster, C. (2002) Property rights and the public realm: gates, green belts, and Gemeinschaft, *Environment and Planning B: Planning and Design*, 29, pp. 397–412.

Webster, C., Glasze, G. & Frantz, K. (2002) The global spread of gated communities, *Environment and Planning B: Planning and Design*, 29, pp. 315–320.

Wilson-Doenges, G. (2000) An exploration of sense of community and fear of crime in gated communities, *Environment and Behavior*, 32, pp. 597–611.

Yelaja, P. (2003) Safe and sound as the iron gates swing shut, *Toronto Star online*, 18 January. Available at http://torontostar.com/NASApp/cs/Content(d = 1035776772468 and call_pageid = 968332188492 (accessed 18 January 2003).

Gated Communities: (Ne)Gating Community Development?

SARAH BLANDY* & DIANE LISTER**

*Centre for Regional, Economic and Social Research, Sheffield Hallam University, Sheffield, UK, **Department of Land Economy, University of Cambridge, Cambridge, UK

(Received October 2003; revised May 2004)

KEY WORDS: Gated community, resident management, leasehold, community, social capital, legal sanctions

Introduction

This paper explores the combined impact of the legal framework for gated communities (GCs) and their physical characteristics that define them as a particular type of housing development. GCs have an explicit boundary, access by non-residents is restricted, the development is usually managed by the residents, and there are legal constraints on residents' behaviour and use of their properties. Three aspects of GCs are addressed here: the extent of social interaction within the walls, attitudes to resident participation in management, and relationships between residents of the GC and of the wider neighbourhood within which it is situated. The first section of the paper sets out the growth of GCs to date, both in England and the US, before moving on to outline the authors' study of an English GC. This was a small-scale pilot study, which nonetheless provides some useful qualitative data on which subsequent discussion is based.

The term 'gated community' (GC) provokes enquiry about exactly what type of community is being referred to; the term is clearly being used to denote a geographical neighbourhood, rather than a non-territorial, liberated community which may be "city-wide, national, international and increasingly virtual" (Forrest & Kearns, 2001, p. 2129). The idea of a GC resonates with many current theories and policy concerns about neighbourhood and community, with a level of governance that could even be seen as an ideal form, in the light of recent UK government policy emphasis on the importance of social capital at a neighbourhood level. Government discourse on social capital provides a useful starting point; three main components are identified in the following extract from a discussion paper:

> *Networks* [neighbours], *Norms* [reciprocity, due care of property, challenging strangers] and *Sanctions* [recognition and respect, versus gossip and social exclusion]. (Performance and Innovation Unit, 2002, p. 11, emphasis added)

Social capital is, of course, a contested concept which has been extensively discussed elsewhere (see, for example, Foley & Edwards, 1999). Entering that debate is beyond the scope of this paper; here it is merely noted that the term is frequently used normatively to denote extensive social interaction at a local level, as well as active involvement in the wider community.

The GC legal framework can be contrasted with the informal social control supposedly exercised within effective communities whose members share collective norms and values. Drawing on findings from research into US GCs, where GCs are long established, this paper explores whether GCs may exemplify the replacement of traditional community and neighbourhood functions with *legal* networks, norms and sanctions. 'Legal networks' refers to the legal framework which ties GC residents together, with collective responsibility for managing the development; the 'norms' are those standards for residents' behaviour and use of their property set out in covenants in the lease; and 'sanctions' refer to the fact that GC residents who fail to observe the legal framework may ultimately have their home repossessed.

The paper now presents an overview of the development of GCs, and their growth and appeal in England; social and legal networks, norms and sanctions will be discussed in subsequent sections of the paper.

The Growth of Gated Communities

The US provides the main context for discussion of English GCs, as the same system of property law applies in both countries, and the growth of GCs there has been well-documented (see, for example, Blakely & Snyder, 1997). In the US there has also been an extraordinary increase over the past 50 years in the number of self-managing neighbourhoods (not necessarily gated). About one in six residents there, around 50 million people, now lives in a community which is governed by a homeowners' association (Rich, 2003). Most English GCs are governed in a similar way, with residents automatically becoming members of the residents' management company on purchase of their property. Ownership of, and responsibility for managing, the common parts of the development (the roads, leisure facilities and communal gardens) is transferred from the developer to the residents' management company. The task of maintaining these facilities is usually carried out by a professional property managing agent, employed initially by the developer and subsequently by the residents' management company when it takes over.

Little is yet known about the 1000 or so gated communities in England, although the national survey of all English planning authorities has provided information about their extent and location (Atkinson *et al.*, 2004). It should be noted here that GCs are not a new phenomenon in this country, either in their built or legal forms. Many housing estates were originally built with private gated roads, for example the 18th century squares of north London, as well as more modest developments in other cities (see Atkins, 1993). The leasehold estate with restrictive covenants provided an efficient way of ensuring early development control in these developments. However, the recent growth in GC developments in this country has stimulated new interest.

Some current data about general attitudes towards GCs in the UK, and therefore future demand, are provided by the findings of a telephone survey carried out on behalf of the Royal Institute of Chartered Surveyors in 2002 (Live Strategy, 2002; for further details see Blandy *et al.*, 2003). A random sample of 1001 respondents throughout the UK was interviewed for this survey. An outline of the findings is included here to provide a context for the subsequent discussions in this paper. It was found that younger people (65 per cent) were more attracted to GCs than older respondents. Half of all respondents said that GCs were a good thing and were in favour of them, although only one-third of respondents agreed with the statement that "living in a gated community appeals to me personally". This latter group included nearly half of respondents living in rented accommodation, compared to slightly less than one-third of the owner occupiers questioned. Respondents on lower incomes were more likely to agree with the statement than those on higher incomes.

All of the 34 per cent (345) respondents who found the idea of living in a GC appealing were asked about which aspects they found most important. The most attractive aspect for those respondents was 'greater security' (72 per cent). This was particularly true of younger age groups. 17 per cent of the 345 cited 'peace and quiet' as the most appealing aspect of gated communities, and 6 per cent 'living with people of similar background'. 'Greater status/prestige' and 'privacy' were each mentioned by 1 per cent of respondents who said they were personally attracted by the idea of living in a gated community. The responses of those who did not find this idea appealing were particularly illuminating for the context of this paper. One-fifth of this group said it would be 'too dull', and half said that they 'would not want to live behind walls or fencing'. Interestingly, 50 per cent of this group (one-third of the total) stated that they 'would rather be part of a community', thus making a distinction between a gated community and an unplanned neighbourhood. This distinction is pursued in the next section of the paper.

The Area and the Gated Community

The empirical data on which this paper is based come from a small-scale pilot study of a GC in Nether Edge, Sheffield. This district is situated between the wealthy south-west sector of Sheffield and the more deprived areas to the east and north. The adjoining ward to the north-east (Sharrow) is one of the most racially mixed areas of Sheffield. The local authority ward of Nether Edge has 17 500 residents and is midway on the national range of indices of deprivation; the neighbouring ward of Broomhill is in the 10 per cent least deprived areas, and Sharrow in the 15 per cent most deprived (DETR, 2000).

Within Nether Edge, there is a range of property types and prices, from a "splendid and substantial Victorian gentleman's residence with six bedrooms" (as described in an estate agent's advertisement) to small terraced houses in less affluent streets, on sale for one-fifth of the price and often used for renting out to students. It is considered locally to be a mixed neighbourhood with a definite and somewhat 'alternative' identity, illustrated by a website entry which refers to the "distinctly Seattle Bohemian feel" of Nether Edge, where the author "sat drinking *mio caffe* ... as a large Pakistani family strolled by" (Mitchell, 2004). There is a thriving Nether Edge Neighbourhood Group which appears to epitomise many of the ideals of social capital, although it is uncertain what proportion of the local population are involved. The Group was founded in 1973:

> with the purpose of preserving and fostering the amenities of the area. ... We publish *EDGE* (the neighbourhood newsletter) nine times a year. Membership is by voluntary subscription. ... Our Planning Group examines every planning application affecting our area ... From its earliest beginnings there has been a luncheon club for the housebound, which provides a midday meal and conversation every Monday, made possible by a team of volunteer cooks and drivers. (*EDGE*, December 2001)

The GC, which will be referred to as Upper Green, was the first of its kind in the city, built on a former hospital site of 4.22 hectares where the first building had been a workhouse in 1841. The development company invested £30 million in the purchase and conversion of the site to an exclusive, yet high density, residential development of which the original surrounding wall has now become the distinguishing feature. The 19th century buildings have been converted to apartments and town houses, and new properties comprising both houses and apartments are being built, to make 180 dwellings in all. There are also communal garden areas in the centre, leisure facilities and a swimming pool which are open only to residents. All three access roads have electronically controlled gates. CCTV cameras are installed around the development set up to automatically track anyone entering through a pedestrian access gate. Once the development of the site has been completed a security guard will be on duty during working hours, and overnight a control centre will take over responsibility for security. At the time of the study, only one-third of the dwellings had been completed.

The developer's solicitors provided sample legal documents for the GC. Each purchaser signs a 200-year lease between the developer, the leaseholders and the residents' management company, and automatically becomes a member shareholder of this management company, signing a form indicating their willingness to stand for the position of director and secretary of the company. When the last plot is sold, the freehold will be transferred to the management company, although for the first year of its operation the developer will retain a 'golden share' to ensure control. The rights of the individual owners are restricted through the use of covenants in the lease, which can be enforced by other individual residents or by the residents' management company on their behalf. The developer has appointed a professional property management agent to carry out ground maintenance and run the leisure facilities; the residents' management company can continue to employ this professional managing agent, or find a substitute.

Research Methods

The developer gave permission for a study of the GC, and their on-site sales agents distributed questionnaires to all new residents over 16 months, from July 2001 to October 2002, during which period the first 60 sales were completed. Twenty-three purchasers completed questionnaires, returned by post in pre-paid envelopes, a response rate of 38 per cent. The four-page questionnaire sought information about why respondents had moved into the GC, where they had moved from, how they envisaged neighbourliness developing, and how important the lease was in restricting their own and other residents' use of their properties. The questionnaire asked respondents to rate suggested responses on a scale of 1 to 5, allowing respondents to add their own responses. Interpretation of the questionnaire responses is qualitative in nature, given the small size of the sample.

Ten respondents to the questionnaire indicated their willingness to be interviewed, and eight semi-structured interviews (including two where both adult residents were present) were carried out. At the time of interview, the residents had been living in the GC for an average of six months. It was therefore possible to compare each resident's original motivation for purchase with their experience of living there, and to probe further about how social relations in the GC and within the wider neighbourhood were developing. Respondents' views were also sought on how any tensions between neighbours might be resolved, and whether the interviewee would be prepared to get involved in the management of the gated community.

The self-selected residents who completed and returned the questionnaire, and those who volunteered to be interviewed, may not be representative of the residents as a whole. Although attempts were made to secure interviews with residents across the age spectrum, three-quarters of the interviews were with residents over 40 years old; it is well established that older people are more likely to express satisfaction on a range of issues. Household composition of interviewee households included a single female and two households with children, the remainder being childless couples. One of the eight interviewee households lived in a newbuild dwelling, whereas one-quarter of the 60 dwellings to which questionnaires were distributed were new. Thus the interviews were mostly with older residents, living in converted houses rather than newbuild apartments. Further, being prepared to take the time to participate in research may indicate a certain public-spiritedness, and might also appeal to people willing to reflect on their experiences as 'pioneers' of a new form of housing; half of the interviews were with the first five questionnaire respondents. The findings from the interviews must be interpreted with these factors in mind.

'Sense of Community' within GCs

Here, the study draws on community psychology literature, which has developed the concept of 'sense of community' as a catalyst in mobilising an individual's perception of their environment, their social relations, and perceived control and empowerment within the community (see Chavis & Wandersman, 1990, p. 56). The Upper Green study provided an opportunity to test the expectations of purchasers about these issues, as well as their experience after some months of residence. In order to determine the relative importance of 'community' for those moving in to Upper Green, the questionnaire asked respondents to rank six factors on a scale of 1–5, with 1 being most important and 5 the least important, in terms of motivating their purchase.

Table 1. Questionnaire responses: purchasers' reasons for moving to the development, in order of importance ($n = 23$)

Reasons for moving (not mutually exclusive)	Very important	Important	Total of first two columns	Neutral	Not important	Not important at all
Property values	69.6% (16)	17.4% (4)	*87% (20)*	4.3% (1)	4.3% (1)	0% (0)
Security features	26.1% (6)	43.5% (10)	*69.6% (16)*	26.1% (6)	4.3% (1)	0% (0)
Leisure facilities	21.7% (5)	39.1% (9)	*60.8% (14)*	26.2% (6)	8.7% (2)	4.3% (1)
Moving into a community	34.8% (8)	17.4% (4)	*52.2% (12)*	21.7% (5)	13% (3)	8.7% (2)

The summary of responses in Table 1 shows that most respondents considered maintenance of property values as most important, followed by security, with good leisure facilities also seen as important. 'Moving into a community' received a more mixed response; it was important for 12 of the respondents, but the other 10 ranked this factor as neutral or unimportant. Other reasons were volunteered by residents for moving into the gated community: 'easy access to city centre', 'a quiet pleasant neighbourhood', 'ease of maintenance', 'investment potential', and the attractive design of the dwellings and of the development as a whole. Taken together, these responses give a picture of the primary motivation of households for purchasing a property at Upper Green, at the point when they had just moved into the development.

The group of respondents willing to be interviewed did not differ significantly from the questionnaire respondents as a whole in the reasons they gave for moving to the GC. However, the subsequent interviews shed more light on the reasons for the preferences expressed in the questionnaire responses and, as might be expected, each household's personal circumstances were reflected in their priority reasons for moving. For example, one family with four children rated all factors (apart from proximity to job) as 'very important' in their questionnaire response. It became apparent during the interview that this female respondent's reasons for moving to a GC were directly related to the fact that her husband worked away, and that she was on her own with the children during the week:

> ... the fact that you're behind the gate—which makes you feel that little bit more secure ... a lot of the time I'm by myself. So the security thing is more sort of important. And I'm not worried then about going out and leaving one of the older children here. (C3).

This household, and another consisting of a single female, both considered the leisure facilities and 'moving into a community' very important. Clearly, for these particular households the GC represented less isolation, and an opportunity to have contact with neighbours when using the leisure facilities. Other respondents found it hard to disentangle the different factors. In an interview a male respondent, whose household consisted of a couple both in their thirties, explained that:

> when you're buying somewhere new, then yeah I definitely expect to have, you know, to have a decent specification of security system ... it was just a kind of,

Table 2. Questionnaire responses: expectations of neighbourliness ($n = 23$)

	Very likely, or likely	Neutral	Not likely, or not likely at all
More contact with neighbours than in an ordinary street	73.9% (17)	21.7% (5)	4.3% (1)
Will make friends through use of the leisure facilities	52.1% (12)	47.8% (11)	0% (0)

add-on factor that ... you know, would add to the whole package, sort of, further on down the line hopefully make it easier to sell on. (B5)

Table 2 presents the responses to the question of how purchasers envisaged 'neighbourliness', or social interaction, developing at the GC. As it is an unusual development, being physically bounded and high density with exclusive access to leisure facilities, the questionnaire asked whether respondents thought that they would have more contact with neighbours in the GC, than in an ordinary street. Nearly three-quarters of respondents thought they would. The respondents were divided on whether friendships might form around use of the leisure facilities.

In the interviews we probed whether respondents' ideas about community and neighbourliness rested on having like-minded, similar neighbours. One interviewee, a single female in her fifties, explained that she felt happy living in the GC because "I feel very safe ... my neighbours are very nice ... they're caring". She then clarified that her contact with neighbours was not as a result of pro-active social activity, but occurred mainly because: "we all park our cars ... sort of like next to each other so you meet people getting out of their cars" and, for her, neighbourliness was manifested by "people just say[ing] hello as you are going past. I don't want to live in everybody's pockets" (F4).

A range of other responses emerged from the interviews about preferences and the potential effects of living within the confines of a gated development. One female interviewee, in her twenties, from a two-person household with no children, said that: "everybody keeps themselves to themselves and just gets on with it, which in a way I prefer". This view of community and neighbourliness was echoed by another female interviewee, from a household of two adults in their fifties with one teenager, who said: "If we found we were getting on top of other people we would move". Another couple explained that they were more attracted to the architectural features of the development and to its situation in the Nether Edge area, rather than it being gated and self-managing, going so far as to disassociate themselves from the GC: "I don't see us being part of the community of this development ... for me, one of the attractions is actually living on the very edge of it" (G22).

A contrasting perspective was provided by a retired male interviewee, from a household of two adults, whose questionnaire response had anticipated much more contact with neighbours than in an ordinary street. In interview this respondent expanded on his views in a rather nostalgic comparison: "we're quite cheek by jowl here, it's almost like living in terraced houses, back to the old days in Sheffield, and it's all very close" (E12).

These findings indicate that property values and security were far more important to most purchasers than 'community'. Only a small proportion of residents were committed to developing social networks within the development. For the majority, a low level of informal associative contact with neighbours was both what they anticipated, and all that they wanted. The primacy of property values as a motivation for purchasing properties in the GC chimes with Forrest & Kearns' suggestion that "people may 'buy into' neighbourhoods as physical environments rather than necessarily anticipate or practice a great degree of local social interaction" (Forrest & Kearns, 2001, p. 2130). Certainly, findings so far indicate that the majority of GC respondents envisaged, and have experienced, a community characterised by weak ties, which nonetheless are important in sustaining identification with the neighbourhood and fellow residents (see Henning & Lieberg, 1996).

Another factor to be taken into account is that this GC is still in development, and changes may well occur in its social character over time. For example, as residents settle in, "... if everyone realises that other residents retain a reserved position as well, this may actually give rise to growing preparedness to meet each other: the 'paradox of reserve'" (Reijndorp *et al.*, 1998, p. 228, translation provided by Reinout Kleinhans). The developer also offered an interesting perspective in interview, on how the layout and facilities of the GC might affect the development of community feeling. He pondered whether "we'll have four distinct communities", based on building types and locations, which might possibly regroup with the opening of the leisure complex in summer 2004.

When management is handed over to the residents' management company, 'keeping to oneself' may become more problematic for both the individual residents and for the effectiveness of self-management. On the other hand, the collective experience of managing the development might draw in more residents. The following section of the paper addresses this issue.

Resident Participation in Management

Government policy has for some years encouraged the participation of social tenants in the management of their estates, in the belief that such involvement builds social networks and capital (DETR/DSS, 2000; SEU, 2000). The literature on tenant participation could be expected to provide some insights into the effects of self-governance on residents and their neighbourhoods. However, its focus is primarily on analysing the development of tenant participation from struggle to mainstream, or on setting out measures which landlords must take to make it effective. With the exception of Birchall (1997), very little has been written about tenants' active participation, and why they participate. The first point to note is that the process of self-governance by residents can be misinterpreted. For example, Birchall described the Cloverhall Tenants management co-operative in Rochdale as 'outstandingly successful'. Yet different researchers found a rather different picture: a few active tenants did most of the work, which caused resentment both amongst the committee members and "a degree of resentment amongst those outside the committee that a small group are in control of the co-op" (Dickson & Robertson, 1993). A study of tenant participation in the social rented sector noted identical themes: general apathy, with few tenants prepared to get involved until motivated and united by a common problem. Concern was also expressed about the motives of tenants who were initially prepared to get involved (Cole *et al.*, 2000).

In the owner-occupied sector in England, residents are not usually expected to participate in managing their neighbourhood, with the exception of residents of blocks of flats held on long leasehold. A study of leaseholders found that the overwhelming reason for pursuing leasehold enfranchisement (acquisition of the freehold) was the behaviour of the freeholder or the managing agent, in other words an *external problem* as in the social rented sector, rather than a desire to collectively manage the jointly owned property. Further, once enfranchised, some residents found that a small group then took control and that genuine power-sharing remained elusive (Cole *et al.*, 1998). A systematic review of empirically-based research into GCs established that all studies had found low social interaction in districts governed by associations (Blandy *et al.*, 2003). One survey of Californian residents' associations showed that few residents participate in management (Barton & Silverman, 1987). Another study found that residents did not expect 'a more zealously formed community' merely because the neighbourhood was run by a residents' association; most assumed that the association's role was limited to physical maintenance of the environment. Further, resident management through the association was irrelevant to the interviewees' decision to purchase. Most expected to remain uninvolved, but could imagine becoming active if conditions deteriorated (Alexander, 1994).

Explanations for this may be found in the literature on participation and community development. Chavis & Wandersman (1990) reviewed previous research in this field and concluded that a 'sense of community' is strongly linked to social control of the neighbourhood. However, their research points to two contradictory forces. They found that in 'problem neighbourhoods' difficult issues can serve as a motivator to collective action and participation. In 'nice neighbourhoods', the more safe and secure residents feel, the more likely they are to interact and feel a sense of community. This is also an incentive to participate, but in what? Without an external problem to motivate residents, there may be no spur to collective action.

In the Upper Green study, the views of residents about how they envisaged the management company being run, and whether they would be prepared to take part in it, were explored in order to assess whether this form of self-governance might enhance the residents' sense of community. The questionnaire did not prompt for a link between these two aspects because the study was interested in how far respondents were conscious of the legal network they had become part of through buying into the GC. Only two respondents were fully aware of the details; and both of these connected the legal form of governance with the development of community spirit within the development. One of these households, a retired couple, indicated on their questionnaire response that they thought 'management of the estate' would be very likely to bring a sense of community; the other anticipated that neighbourliness would be enhanced by being 'part of the residents' association'. However, the majority of the residents interviewed had not previously understood that they would be called upon to manage the community themselves when the last plot was sold.

This lack of knowledge was anticipated from the American research literature, and because of the sheer complexity of the Upper Green legal documents, which include a 23-page lease with seven schedules as well as the management company documents. Five of the eight interviewees were confused about the difference between the residents' management company, and the professional management company appointed by the developer, which had already started on maintenance work around the site. They were not clear about their own role, nor about when or how the officers of the residents' management company would be elected. This resident expressed a typically muddled, and passive, view of the process: "we're not sure about ... how that will materialise. We're assuming it will be facilitated somehow" (C3).

The developer's views on the issue of residents' management companies were sought in interview. Based on his previous experience of similar developments, he reflected remarkably similar themes to those that emerge from research into tenant participation:

> ... we [the developers] are always a very positive, uniting, force—in the negative. ... But, I just think communities just form. There'll be those who don't care, there'll be those who care as long as somebody else does it, and ... there's always one or two who always wanted their opportunity to be a local politician. ... you tend to watch how it's beginning to form, and then we cultivate the right people to lead it, which is in our best interests, but also hopefully in the residents' best interests.

Once they realised what the management framework involved (having been told by the interviewer), more than half of the interviewees said they would consider giving some time to the management of the GC. However, echoing the tenant participation literature, and the developer's own experience, some said that time constraints would prevent them getting involved, and one respondent was motivated by an already emerging dispute:

> It is something that I would probably consider because there are certain things that people have suggested [speed bumps] that I think are totally unreasonable, so I want to make sure that those people aren't the ones that are going on [to the committee]. (A1)

It is clear from this brief review that residents, across all tenure sectors, are often motivated by a perceived external threat, and also regardless of tenure that a small group can take control. In American suburbs governed by homeowners' associations, apathy is a more common problem than resentment about the power exercised by board members. The qualitative data from the Upper Green study indicate that a legal form which requires residents to collectively manage their community is no guarantee of active participation, particularly as most residents interviewed were simply unaware of the legal framework through which they would take responsibility for running the development.

Enforcement of Behaviour through Law: Norms and Sanctions

In a GC such as Upper Green, standards of behaviour are set out in the lease, and legal sanctions are available to enforce these. We are not suggesting that this makes it impossible for social norms and sanctions to be developed alongside the legal framework, but that it is interesting to explore how social networks may develop against this background. In the UK, some policy makers advocate tying residents of disadvantaged neighbourhoods into a legal agreement to make acceptable behaviour standards crystal clear. This guidance is invariably directed towards tenants, but the Upper Green covenants also aim to control the occupiers' behaviour and use of their property, in a way which seems at odds with our expectations of the freedoms enjoyed by owner occupiers. For example, the lease forbids children's play in any communal areas except the designated play area; it also restricts the use of leisure facilities to those who permanently reside in the development, which clearly envisages a very self-contained community. Further clauses require residents to clean their windows at least once in every four weeks; not to hang or spread any laundry or other paper anywhere outside; and not to place any pots or other papers on any exterior window sill. There are also standard covenants against nuisance and anti-social behaviour, including excessive noise between 11.00pm and 8.00am.

These legal standards can be compared with the social norms of 'reciprocity, due care of property, challenging strangers' (Performance and Innovation Unit, 2002, p. 11). It is hard to envisage how reciprocity could be enforced through law, but the covenants in the Upper Green lease should ensure due care of property. The function of challenging strangers is taken on by the electronic gates rather than the residents. A further question is how far the new purchasers had absorbed the details of the covenants. Most of the interviewees were able to recall one or more of the covenants, sometimes incorrectly, giving the impression that they did not have a thorough knowledge of the detailed provisions. This was unsurprising, as a previous US study found that less than 10 per cent of GC residents had read the covenants prior to purchase (Alexander, 1994). However, at Upper Green, one respondent had been through the lease with a fine-tooth comb, and had negotiated various changes such as permission to erect a rotary dryer in the garden. Another household was well aware of the covenants, having initially been sent the wrong lease.

Respondents to the questionnaire were asked to assess the importance of the covenants (see Table 3). Their responses indicate that most residents did not accord any importance to the fact that the covenants restricted their own behaviour. However, a large majority of respondents felt that it was very important that the restrictions would ensure that all other residents kept to the terms of the lease. The eight interviewee households also reflected these views, with two of them suggesting that the covenants were a positive feature of the GC, ensuring uniformity of the look of the development as well as conformity to acceptable standards of behaviour. This dualistic view of the covenants indicates that, even in a development whose high property values could be seen as a proxy for sharing similar social norms, respondents felt reassured that the covenants would ensure appropriate behaviour by other residents.

Turning to sanctions, the important factor to note here is that, because the development is managed collectively by the residents, they are responsible for enforcing the covenants enshrined in their legal documents. It is known that in the US, disputes over enforcement of covenants, conditions and restrictions are common (see for example, McKenzie, 1994; Rich, 2003) with 16 per cent of homeowners' associations using the small claims court and 12 per cent the municipal court. Forty-one per cent of Californian associations reported

Table 3. Purchasers' view of the leasehold covenants ($n = 23$)

	Very important, or important	Neutral	Not important, or not important at all
Assessment of importance in restricting respondents' use of own property	30.4% (7)	30.4% (7)	39.1% (9)
Assessment of importance in ensuring all residents comply with the terms of the lease	78.2% (18)	21.7% (5)	0% (0)

suffering from major problems with rule violations, and 44 per cent of board presidents had been personally harassed, threatened with litigation or actually sued by a member in the previous year (Barton & Silverman, 1987). Current data are hard to find, but several US states are now introducing legislation to regulate the activities of homeowners' associations, and analysis of court records in the Houston area alone identified more than 15 500 cases filed by associations between 1985 and 2001 which could have led to foreclosure and eviction (see Rich, 2003).

The Upper Green interviewees were asked how they would deal with a neighbour who was clearly in breach of their covenants. One, who had already had to sort out a dispute over common garden areas with a neighbour, gave a typical response:

> Well, we dealt with it by having a conversation really, as you would anyway. But I think anything major ... would be for the management committee. (C3)

In a relatively small self-managed development like Upper Green, the members and officers of the residents' management company would be known to all residents. The enforcer of covenants would not be an outsider, such as the landlord in the rented sector, but possibly the near neighbour of the offending resident. This may explain why feelings run so high over breaches of CCRs in US homeowners' associations. At Upper Green, a strong legal regime and enforcement through the management committee means that residents do not need to negotiate shared standards of behaviour, or enforce these through the social sanctions of 'gossip and social exclusion' (Performance and Innovation Unit, 2002, p. 11).

Relationship with the Wider Community of Nether Edge

One concern often expressed about GCs is that their exclusiveness may adversely affect the neighbouring community, yet there is no empirical evidence on this issue. In this study, no interviews have been carried out with local residents who live outside the GC. Here, the research draws on interviews with the GC residents, and on papers and letters in the newsletter of the Nether Edge Neighbourhood Group, to explore possible tensions between those inside and those outside the gates. The development of the GC attracted a great deal of attention, and residents of Nether Edge have used the newsletter to express and share their (overwhelmingly negative) reactions to the GC as it was planned, marketed, and built:

> What appalled me about the brochure were repeated references to SECURITY. Has no one told [the developers] and their prospective purchasers that Nether Edge is already a safe and pleasant place to live? The notion that strangers equal trouble is not only mistaken but also deeply offensive. (Article, *EDGE*, October 2001)

Two themes have emerged from the pages of *EDGE*: the perceived social withdrawal of the Upper Green residents, and the portrayal of Nether Edge as an area of social mix and inclusion. The following is typical of the strong feelings of nearby residents:

> By shutting themselves in, and thereby excluding us local 'undesirables', they have failed to realise that life in Nether Edge is also about people; about sharing and caring; about the rich variety of culture in our local community, the inclusion of

> those who have different values and beliefs. Inclusion will not make life more insecure, exactly the reverse. (Letter, *EDGE*, May/June 2003)

In interviews with GC residents, it was asked whether the respondent household would join the Neighbourhood Group. Interestingly, the two households which were identified as feeling uncomfortable with the security features of the GC, were both very positive about becoming members of NENG, including one respondent whose completed questionnaire stated that "contact with the wider Nether Edge community" would increase the sense of neighbourliness. However, other residents reacted to the hostility of some local people, as illustrated by this comment:

> I've had the little Newsletter thing through which I thought was quite interesting because they spend the whole article saying how it was like Colditz, even though now they've been up having a good snoop round, which I thought, well ... I don't really want to be part of them if that's what they think. (A1)

Another gated community resident referred to some graffiti:

> just on a wall, some idiot had daubed, 'This way to the middle class ghetto' ... even old neighbours of ours down there [referring to their previous house in another part of Nether Edge] said, "Oh, are you moving to that ghetto?" sort of thing. (E12)

Nether Edge may be an unusual area in that many of its residents are extremely positive about its social and ethnic mix, and are actively socially engaged. However, the reaction of its residents indicates the problems that may emerge when a 'ready-made' GC development is created in the midst of an established neighbourhood, raising important issues about exclusion and exclusivity.

Conclusions

This paper has considered the impact that GCs have on the residents of the wider community and on those within the confines of the walls. With a focus on legal relations, it has explored GC residents' sense of community and neighbourliness, as well as their participation in the management of the development. Overall, this review of the research literature and the findings from the Upper Green study suggests that the legal framework of GCs does not enhance social capital, in the sense of social networks, norms and sanctions. Residents of GCs tend to have weak social ties. At the time of the decision to purchase, the majority are simply not aware of the implications for collective management by the residents. Future participation can be expected to follow a similar pattern as in other tenures, where a passive majority allows an active minority to take over running the development.

The covenants in GCs' legal documents can be seen as a substitute for shared standards of behaviour negotiated by neighbours; instead, concerns about standards of behaviour amongst the GC residents are addressed in a legal form. When breaches of covenant occur, as they inevitably will, self-management by residents will be put to the test. Perhaps residents of English GCs will not resort to litigation to such an extent as their American counterparts, but once neighbour relations are legalised it becomes logical to involve the law in dealing with any contravention.

Finally, in this study, relations with the wider neighbourhood seem to be adversely affected by the physical form of the GC development. This has far-reaching implications for community cohesion. There is a danger of a 'them and us' attitude developing both amongst residents of the GC, and of the surrounding neighbourhood. In the Upper Green study, this antagonism may be particularly marked because of the nature of the area in which the GC is situated. The empirical data used in this paper, although from a small-scale study, nevertheless highlight the need for further research to inform policy, so that the longer-term social effects of GCs can be more fully considered.

References

Alexander, G. S. (1994) Conditions of 'voice': passivity, disappointment and democracy in homeowner associations, in: S. E. Barton & C. J. Silverman (Eds) *Common Interest Communities: Private Governments and the Public Interest* (Berkeley: Institute of Governmental Studies Press, University of California).

Atkins, P. (1993) How the West was won, *Journal of Historic Geography*, 19, p. 3.

Atkinson, R., Blandy, S., Flint, J. & Lister, D. (2004) *Gated Communities in England*, New Horizons research series (London: ODPM).

Barton, S. E. & Silverman, C. J. (1987) *Common Interest Homeowners' Associations Management Study* (Sacramento: California Department of Real Estate).

Birchall, J. (1997) The psychology of participation, in: M. Hawtin & C. Cooper (Eds) *Housing, Community and Conflict* (London: Ashgate).

Blakely, E. J. & Snyder, M. G. (1997) *Fortress America: Gated Communities in the United States* (Washington DC; Cambridge, MA: Brookings Institution Press; Lincoln Institute of Land Policy).

Blandy, S., Lister, D., Atkinson, R. & Flint, J. (2003) *Gated Communities: A Systematic Review of the Research Evidence*, CNR Paper 12. Available at http://www.neighbourhoodcentre.org.uk/research/cnrpaperspdf/cnr12pap.pdf.

Chavis, D. & Wandersman, A. (1990) Sense of community in the urban environment: a catalyst for participation and community development, *American Journal of Community Psychology*, 18, pp. 55–81.

Cole, I., Gidley, G., Robinson, D. & Smith, Y. (1998) *The Impact of Leasehold Reform. Flat Dwellers' Experiences of Leasehold Enfranchisement and Lease Renewal* (London: DETR).

Cole, I., Hickman, P., Millward, L., Reid, B., Slocombe, L. & Whittle, S. (2000) *Tenant Participation in England: A Stocktake of Activity in the Local Authority Sector* (Sheffield: Centre for Regional, Economic and Social Research, Sheffield Hallam University).

DETR (Department for the Environment, Transport and the Regions) (2000) *Indices of Deprivation 2000, Regeneration Research Summary Number 31* (London: DETR).

DETR/DSS (Department for the Environment, Transport and the Regions / Department of Social Security) (2000) *Quality and Choice: A Decent Home for All* (London: ODPM).

Dickson, J. & Robertson, J. (1993) *Taking Charge* (London: the London Housing Unit).

Foley, M. W. & Edwards, B. (1999) Is it time to disinvest in social capital?, *Journal of Public Policy*, 19, pp. 141–173.

Forrest, R. & Kearns, A. (2001) Social cohesion, social capital and the neighbourhood, *Urban Studies*, 38, pp. 2125–2143.

Henning, C. & Lieberg, M. (1996) Strong ties or weak ties? Neighbourhood networks in a new perspective, *Scandinavian Housing and Planning Research*, 13, pp. 3–26.

Live Strategy (2002) *Telephone Survey into attitudes towards gated communities in England*, for the Royal Institute of Chartered Surveyors. Unpublished data.

McKenzie, E. (1994) *Privatopia: Homeowner Associations and the Rise of Residential Private Government* (New Haven and London: Yale University Press).

Mitchell, J. C. (2004) *Double Shot Buzz* (website entry for 30 April 2002). Available at http://www.coffeebeer.co.uk/doubleshot/cafe9_toothpaste.html (accessed 3 September 2003).

Performance and Innovation Unit (2002) *Social Capital: A Discussion Paper*. (Available on Cabinet Office website: www.cabinet-office.gov.uk/innovation/2001/futures/attachments/socialcapital.pdf (accessed 3 September 2003).

Rejindorp, A., Kompier, V., Metaal, S., Noi, I. & Truijens, B. (1998) *Buitenwijk: Stedelijkheid op afstand (Suburban Neighbourhood: Urbanity at a Distance)* (Rotterdam: NAI Uitgevers).
Rich, M. (2003) Homeowner boards blur line of who rules roost, *New York Times*, 27 July, p. 1.
SEU (Social Exclusion Unit) (2000) *National Framework for Neighbourhood Renewal: A Framework for Consultation* (London: Cabinet Office).

Who Segregates Whom? The Analysis of a Gated Community in Mendoza, Argentina

SONIA ROITMAN
Development Planning Unit, University College London, UK

(Received October 2003; revised May 2004)

KEY WORDS: Gated communities, urban social segregation, Argentina

> The central problem of our societies is the division among people, and that division is increasingly reflected by walls dividing them, walls whose social weight and impact has increasingly overshadowed their physical might. (Marcuse, 1994, p. 41)

Introduction

In recent years numerous papers, articles and books on gated communities have been published. Many of them emphasise the idea that the arrival of gated communities is closely related to urban social segregation. However, in many cases this theoretical assumption is not validated by empirical data and this paper shows evidence from a case study to support this idea. In the sense that gated communities contribute to the segregationist tendencies that characterise the urban dynamic, the causes and consequences of urban social segregation are discussed in this paper. The structuration theory of Giddens, in which agency and structure influence each other overcoming this duality, is suggested as a theoretical framework to analyse the process of urban social segregation in relation to gated communities. Therefore, structural as well as subjective causes need to be analysed together to understand this phenomenon. On the other hand, it is relevant to address the issue of who is the segregator and who is the segregated as both sides feel segregated. To this end, it is important that the perceptions and opinions of both the residents of the gated community and those of the surrounding community, chosen as the case study, are analysed.

There is a vast literature on the phenomenon of gated communities in Buenos Aires, but there are few works concerning intermediate cities in Argentina. One of the aims of this case study from Mendoza is to make a contribution towards filling this void.

The first part of this paper concerns gated communities and the causes and consequences of their arrival, and the specific conditions in Argentina that favoured this process during recent years. Considering that there are two groups of causes, subjective and structural, but that they have to be considered together, the structuration theory of Giddens is suggested as a theoretical tool to analyse this process.

The second part of the paper is about urban social segregation as a feature of the city. There is a debate in the literature regarding the advantages and disadvantages of segregation. This is discussed in order to press the argument that the effects of social segregation are more negative than positive.

The third part presents a case study of a gated community in Mendoza, Argentina, called 'Palmares'. The final section draws together the main points discussed in the paper and puts forward a possible answer to the question posed in its title.

Gated Communities: Causes and Consequences of their Arrival

In recent years gated communities have proliferated world-wide. They can be defined as closed urban residential areas where public space has been legally privatised, restricting access. They include private property, individual houses and collectively used common private property, for example clubhouse and sports facilities. They have security devices such as walls, fences, gates, barriers, alarms, guards and CCTV cameras. By and large, the infrastructure and services are of a high quality. They are designed with the intention of providing security to their residents and prevent penetration by non-residents, being conceived as closed places since their inception. Law reinforces their closure as private places, which distinguishes them from other places in the city. Their residents must follow a code of conduct concerning social behaviour and construction regulations. With regard to their management, gated communities usually have a residents' association that runs the administration of the neighbourhood and establishes and enforces rules. Gated communities appear as homogeneous places in comparison to the heterogeneity of the 'open city'. Most of their residents are upper- and middle-class families. Laws and regulations, in addition to price of the land and houses, underpin this homogeneity.

There are different causes for the arrival of gated communities, which can be divided into two groups: structural and subjective. The former are influenced by the social, political and economic structure, while the latter are a result of the motives and desires of the social actors. As both of them influence each other, they have to be considered together. Within the first group the most important causes are the rise of insecurity and fear of crime, the deficiency of the state in providing basic services to citizens, increasing social inequalities, the advancing process of social polarisation, as well as an international trend encouraged by developers.

Urban violence and fear of crime are mentioned in the literature as the main reasons for moving to a gated community (see, for example, Blakely & Snyder, 1997; Cabrales Barajas & Canosa Zamora, 2002; Caldeira, 2000; Carvalho *et al.*, 1997; Landman, 2002; Low, 2000; Rovira Pinto, 2002; Svampa, 2001). This is strongly related to the rise in the crime rate. In the case of Argentina, the data show that between 1980 and 2000 the crime rate increased 376 per cent.[1] Considering the crisis concerning the public budgets in recent years, the state has decreased the services provided to the citizens. In the case of housing in the province of Mendoza, as an example, in the year 1991 the state built 4209 houses for low-income families, while in 2001 only 2958 houses were built.[2] The increase in private security hired by companies, banks, shopping malls as well as residential developments is another indicator of the failure of the state to provide enough public police to patrol and control the city. Unfortunately there are no data available about this. However, the public demonstrations held in many Argentinan cities in recent years demanding more security are another indicator of the incapability of the state to cope with violence in the city.

With regard to the income level of the population, the gap between the affluent and the poor is increasing, exacerbating the process of social polarisation. In 2003 in Argentina, the poorest 10 per cent of the urban population in the country earned 2.2 per cent of the total income, while the richest 10 per cent earned 30.8 per cent of the total income. In 1980 the richest 10 per cent of the population earned 15 times more than the poorest 10 per cent, while in 2003 they earned 24 times more.[3] Moreover, the arrival of gated communities has been encouraged by developers as an international trend. This is related to the arrival of foreign investments that want to export models from other places, particularly the US, to developing countries (see Ciccolella, 1999; Thuillier, 2000).

Among the subjective reasons for the arrival of gated communities, the most relevant ones are those based on the desires and expectations of families to achieve a better lifestyle; the avoidance of city problems, such as people asking for money and food; and the search for social homogeneity, status and exclusivity within some social groups in the context of a general process of impoverishment of the society. As an example, the appearance of gated communities in Mendoza has to be considered within the context of the economic crisis that Argentina has faced during recent years, which has led to a general impoverishment of society. Nowadays, while gated communities are increasing in Mendoza, 53 per cent of its population lives below the poverty line (DEIE).

Many authors have referred to the process of choosing a gated community as an act of voluntary segregation, a conscious act and decision taken by an individual or family, contributing to the process of urban social segregation (Borsdorf, 2002; Greenstein *et al.*, 2000; Ickx, 2002; Janoschka, 2002). Other authors say that gated communities represent a special type of segregation (Arizaga, 2000; Carvalho *et al.*, 1997; Marcuse, 2001; Prévot Schapira, 2000). However, the process of urban residential segregation to which gated communities contribute should not be considered as voluntary without considering all the causes that have an influence in its development. This leads to a debate about the room for action that people have regarding the decision of residential location.

While there is no specific theoretical framework to analyse the phenomenon of the gated communities, the existence of structural and subjective causes of their arrival can be related to the main ideas of the structuration theory of Giddens (1984). He argues that the social system is made up of conditions that bind the action of the agent, but do not determine the agent's activity. The structure, which consists of rules and resources recursively implicated in social reproduction, is thus the medium whereby the social system affects individual action as well as the medium whereby individual action affects the social system. While the agent is a knowledgeable actor who has room for action and can influence society, the conditions of society also influence her/his movements. Consequently, the actor is influenced by increasing rates of crime, increasing social inequalities and the inefficiency of the state to provide basic services. However, she/he has the ability to choose whether to move to a gated community or not according to her/his interests, desires and expectations; and also to her/his financial situation that allows her/him to afford living in a gated community or not. Consequently, gated communities contribute to a type of segregation that cannot be defined as either voluntary or subjective, but rather influenced by both.

Gated communities have a specific physical impact upon the urban built environment, such as the closure of streets, the hindrance of emergency services and the fragmentation of the space. In addition, they have political impacts as they arguably undermine the concepts of democracy and citizenship and weaken the role of the state. Furthermore, they have social impacts, such as their role in the process of urban social segregation that influences social life and especially social relations. In this context the question of who segregates whom appears. To have a better understanding of the meaning of this question, the concept of urban social segregation in relation to the arrival of gated communities as well as the debate about its effects need to be discussed. The next section deals with this issue.

Urban Social Segregation: A Debate on its Effects

The dichotomy integration-segregation is one of the most important features of the urban space. The city is a social entity that integrates people through the development of social practices in everyday life such as the use of public spaces, use of public transportation, use of common services such as health services, educational services, recreational services and the provision of work. In opposition to this, the social system tends towards social segregation. People tend to concentrate according to their similar characteristics. These two tendencies coexist in the urban space. Gated communities are understood as a contribution to the segregationist tendencies in the city.

Urban social segregation can be defined as a social process that results in the detachment of certain individuals or social groups, kept isolated by a limited or non-existent interaction with the rest of the society or with other social groups. In the case of the gated communities the detachment and separation is based mostly on income level and residents' desires. The difference between groups is highlighted by the use of physical barriers like fences, walls, barriers as well as other security devices such as guards, dogs and CCTV cameras.

There is a debate in the literature about whether urban social segregation is a positive or negative phenomenon or whether it has advantages or disadvantages. According to some scholars (Blauw, 1991; Greenstein *et al.*, 2000), it might be considered a good phenomenon as it preserves customs and lifestyles and strengthens social and identity ties through the development of nets of reciprocity and help. Furthermore, in the case of minority groups, spatial concentration can provide a better position to defend their interests and have more power, electoral influence being one obvious example.

In the case of the gated communities the process of urban social segregation might be characterised as positive since it allows their residents to reinforce social homogeneity and sense of community and to protect themselves from what they perceive as the danger and violence of the 'open city'. However, the idea of the development of a sense of community in this type of neighbourhood has been questioned. Researchers such as Wilson-Doenges (2000) have showed that gated communities are not such a safe place to live and there is no sense of community different from the one that could be in the 'open city'. Moreover, there is no guarantee that they will repel more thieves than they attract. Burglars know that the doors and garages are likely to be unlocked and houses are likely to have valuable objects inside, so once the main wall has been passed, there are no difficult obstacles to overcome.

Despite these positive effects, according to the literature the negative impacts of segregation are more evident and dangerous in terms of society as a whole. Social segregation might lead to feelings of exclusion and being rootless, and worsen problems of social disintegration. Moreover, segregation has effects on employment, as there are fewer possibilities to get information about job opportunities. There is a large body of research that links segregation with poverty and the underclass (the 'ghetto effect') (see Massey & Denton, 1993; Mingione, 1996; Wilson, 1987). In addition, poor people who live segregated lives face less chances for upward social mobility. Following this argument, Blakely & Snyder (1997) argue that segregation has a variety of negative impacts, such as reduced opportunities, concentration of deprivation, greater vulnerability to economic downturns and separation and isolation not just from other members of society, but often also from jobs, adequate public services and good schools.

On the other hand, it is important to note that although most of the time the bad effects of living in a segregated place, which is not only segregated but poor as well, are highlighted, living in a segregated but wealthy place like a gated community also has many drawbacks for its residents. Social segregation hardens and breaks the social fabric through the use of visible barriers that do not allow strangers to go inside the borders of the gated communities. It reinforces social differences and social divisions. There is a lack of contact with different people. The construction of social relations is influenced by the separation established between 'the insiders' (the residents of the gated communities) and 'the outsiders' (chiefly the surrounding community). 'The others', who are the people outside and especially the neighbours in the surrounding areas, are perceived by the residents of the gated communities as strangers and as potential aggressors. In this way the physical barriers are used to establish a distance, which is not only physical, but also social and symbolic.

In addition, there is a close relation between segregation and social exclusion, as the former might conduct towards the latter. Castells (1998) defines social exclusion

> ... as the process by which certain individuals and groups are systematically barred from access to positions that would enable them to an autonomous[4] livelihood within the social standards framed by institutions and values in a given context; ... in informational capitalism, such a position is usually associated with the possibility of access to relatively regular, paid labour, for at least one member of a stable household. (p. 73)

The literature on gated communities emphasises the idea that this type of residential development contributes to urban social segregation (e.g. Blakely & Snyder, 1997; Caldeira, 2000; Low, 2003; Svampa, 2001). However, in most of the cases there is not enough empirical data to corroborate this statement.

The next section is about the gated community chosen for the case study. The data collected during the qualitative research illustrate the feelings, perceptions and opinions of the residents of this gated community as well as of the surrounding community concerning the link between gated communities and urban social segregation.

Case Study: A Gated Community in Mendoza, Argentina

This section is about the case study chosen in order to carry out the qualitative research to support the theoretical ideas discussed above. The first part is about the context in which the research was conducted, emphasising the main features of Mendoza and Argentina that allow us to understand the phenomenon of the arrival of gated communities. The following part describes Palmares, which is the gated community chosen for the case study. There then follow descriptions of the methodology and the research findings.

The Context: Argentina and the Metropolitan Area of Mendoza

The expansion of the gated communities during the 1990s in Buenos Aires is one of the most important transformations of urban landscape in the country. Gated communities in Argentina are inhabited by upper-middle- and upper-class families. This process of gating has occurred later than in other Latin American countries or in the US (Svampa, 2001). In the beginning it was more a consequence of a strategy of distinction, while in recent years it is related to a new logic of urbanisation and increasing insecurity.

There are more than 400 developments (Svampa, 2001) with about 50 000 residents in Buenos Aires. In the case of the intermediate Argentinian cities, the impact of this residential development has not yet been so radical (see Bragos *et al.*, 2002 about gated communities in Rosario, Argentina, where there are seven closed neighbourhoods). Their development post-dated that in Buenos Aires. Unfortunately there are few studies on gated communities in intermediate cities in Argentina. However, gated communities have had a very visible impact on many cities as they represent a new urban phenomenon.

In the case of the Metropolitan Area of Mendoza, there are more than 45 gated communities, but no data exist about the population living there. Most of them are on a small scale, however, there are two that are considered of significant size as more than 400 families live in each of them. One of these is called Palmares and it is the gated community chosen as the case study for this research.

The province of Mendoza is situated in the west-centre of Argentina, bordering Chile. On the basis of population, it is the fourth largest province in the country, with 1 579 651 inhabitants in 2001, according to the last national census (DEIE). The biggest population is the Metropolitan Area of Mendoza with 986 341 inhabitants (INDEC), which is 62 per cent of the total population of the province and for which it is considered an intermediate city.

Since the 1970s, but particularly during the 1990s, the Argentinian economy has been dominated by neo-liberal policies such as privatisation, decentralisation, withdrawal of many subsidies and a more deregulated, open economy. The opening of the economy to direct foreign investments in the country has led to big changes in the spatial configuration of the city with the appearance of shopping malls, international hotels, multiplex cinemas and gated communities. The construction industry recovered in the 1990s after a long period of crisis due to hyperinflation. In addition, regulations regarding urban planning became more flexible. New investments have produced more segregation in cities. They are very selective regarding their location, which creates highly developed areas that are continuously refurbished and improved and areas that are just forgotten by investors.

Argentina was known in Latin America for having a strong middle-class population. However, the neo-liberal model led to a process of weakening of the middle-class and increasing polarisation. Salaries decreased and since 1995 the rate of unemployment in Argentina has exceeded 10 per cent, reaching 20 per cent in some localities. The rate of unemployment in the whole country in May 2003 was 15.6 per cent and coincidentally, in Mendoza in May 2003 it was the same, 15.6 per cent (INDEC; DEIE). Most of the population became poorer while a select few enjoyed the benefits of the neo-liberal economic model. Although the middle-class is still a very important social group in the Argentinian social structure, the number of people living in poverty is constantly increasing. In May 2003, 42.6 per cent of urban households in Argentina were living below the poverty line[5] according to INDEC, while three years ago, in October 2001, this figure was 28.0 per cent. These recent fluctuations in the economic fortunes of the Argentinian middle-class have probably been more extreme than in the rest of Latin America.

In the Metropolitan Area of Mendoza, in 1990 the poorest 30 per cent of the social structure received 13.7 per cent of the total income, while the most affluent 10 per cent received 25.4 per cent. This discrepancy was getting wider and in 2002 the poorest 30 per cent received 10.7 per cent of the total income and the richest 10 per cent received 27.6 per cent. Hence, in 1990 the richest 10 per cent of the population earned 14 times more than the poorest 10 per cent, while by 2002 that figure had increased to 25 times (DEIE).

It is also the case that the crime rate has increased in the last decade in Argentina and especially in Mendoza. In the former there were 149.8 crimes for every 10 000 people in 1991; by 2000, this had increased to 305.1 cases of crime for every 10 000 people (INDEC). In Mendoza, there were 199.4 crimes for every 10 000 people in 1991, and in 2000, this increased to 570.3 cases for every 10 000 people (a figure exceeded only by Capital Federal). The dominant form of crime is against property (burglaries). As a consequence, all kinds of security strategies have been developed, with gated communities being one of the most radical examples.

The Case Study: Palmares

Palmares was selected as a case study because of its size, age and prestige within Mendocinean society. It is known by everybody, chiefly due to the existence of a shopping mall called 'Palmares Open Mall', which is next to the gated community and was built by the same developer company, where many people go for leisure and shopping. The neighbours in the surrounding areas are middle-, lower-middle- and lower-class families. Social differences are evident in the area as there are notable contrasts in the way people live within the walls of Palmares and the way families live in the surrounding areas, particularly regarding security measures. One of the most important problems that

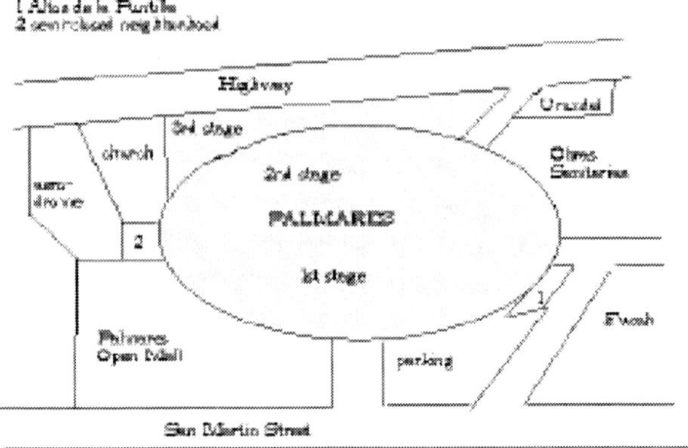

Figure 1. Map of Palmares and the surrounding area (drawn by the author) (not to scale).

the Metropolitan Area of Mendoza is currently facing is the increase in the crime rate and, consequently, more fear of crime. All kinds of security measures are developed such as fences and alarms. In recent months residents of some middle-class and poor neighbourhoods asked the authorities for permission to close their neighbourhood in order to be protected from burglaries and crimes.

The development of the gated community 'Palmares de Presidente' (referred to as Palmares elsewhere in the paper) started in 1993. It is located in the Municipality of Godoy Cruz, in the Metropolitan Area of Mendoza. Palmares, which means 'palm trees', is situated next to a highway that connects the west side of the Metropolitan Area of Mendoza. It is surrounded by a shopping mall, a semi-closed middle-class neighbourhood, a church, a poor neighbourhood (social housing) crossing the highway and, middle- and lower-middle-class neighbourhoods as well as a slum on the north side (Figures 1 and 2).

The gated community covers an area of 55 hectares. There are approximately 700 plots aggregating the three stages of the project's development. The average size of a plot

Figure 2. View of the gated community from the highway side. *Source:* All pictures were taken by Sonia Roitman with authorisation (not to be used without permission from the author).

Figure 3. One of the main roads in Palmares.

is 420 m^2. Some residents have bought more than one plot. The first stage of Palmares started in 1993; 285 plots all sold out and houses were soon being built. The second stage, which has more than 300 plots, was also very successful. The plots of the third (and last) stage, approximately 120, went on sale in December 2002 and half of them have been sold. In April 2003 there were 259 occupied houses dating from the first two phases, 132 houses under construction, and a further 198 empty plots (information given by the Residents' Association of Palmares in April 2003). (Figures 3 and 4).

The price of the land has increased strikingly, from US$3 per m^2 when the land was not urbanised to US$110 per m^2. (In Argentina, until December 2001 due to the Convertibility Law (Ley de Convertibilidad) $1 = U$S1.) In April 2004 the price was US$110 per m^2 in the first and second stages and US$75 in the third stage. The third stage is cheaper than the others as it is so close to the highway (Corredor del Oeste) and it has recently been urbanised (land values provided by a real estate company from Mendoza). The price of the land was accessible because the plots were not so large, but the price of the land per m^2

Figure 4. A street of the first stage and at the back a house under construction in the second stage.

Figure 5. One of the most luxurious houses, situated in a strategic location and is easily seen.

was expensive, being one of the highest prices in the housing market in the Metropolitan Area of Mendoza.

In an attempt to characterise the residents of Palmares it could be said that most of them are in their thirties and forties and have children. There is another group of residents in their fifties with teenagers and young sons and daughters, or who live alone because their children have left home. Most of the residents are upper-middle class who work in the private sector as managers, lawyers and accountants or are the owners of important shops or companies. There is also a group of nouveau-riche residents, distinguished from the rest by their enormous and luxurious houses[6] (Figure 5). Although they are not the richest of Mendoza (with a few exceptions), they are all within the group of the 10 per cent richest due to the economic crisis and the devaluation of salaries.

Methodology

The research used a qualitative methodology. Semi-structured interviews were used to collect the data and 94 interviews were carried out between January and April 2003. There are three different groups of interviewees: (1) the residents of Palmares and people who work there; (2) the neighbours of the surrounding areas outside Palmares; (3) policy makers, developers and researchers.

For the purpose of this paper, two distinct main groups are considered: the residents of the gated community and the neighbours who live outside in the surrounding areas. Within the first group, 48 people aged 13 to 80 years of age were interviewed, 25 were female and 23 male. Of those interviewed, 4 were members of the board of the residents' association that runs the management of the gated community. The interviewees were chosen by snowballing, with the exception of the members of the board, who were purposively chosen. The need to keep a balance between the different social groups in terms of age, time of living in the neighbourhood, friendship, location of the house and size of the house was a priority during the research.

The outside community is a very heterogeneous group and 19 people were interviewed using a snowballing technique. However, in each of the outside neighbourhoods the president or a member of the board of the residents' association was purposively chosen considering the relevance of his/her opinion and knowledge of the relation with the gated community. Within the 19 interviewees, 5 lived in the slum, 10 lived in the two middle-class neighbourhoods, and 4 in the houses next to Palmares, which were built by the same company that developed Palmares and have similar construction features to those in Palmares, although the houses and plots are of a smaller size and of less quality.

Research Findings

In this section the theoretical issues mentioned in the previous sections are supported by the empirical data collected during the fieldwork. First, the reasons for moving to a gated community are discussed, according to the interviewees. Next, the feelings and perceptions of the two groups are mentioned and finally, issues about discrimination and segregation are raised according to the interviewees' experiences. The final aim is to arrive at an answer to the main question of this paper: who segregates whom?

Concerning the reasons for moving to a gated community, as the statistics of Mendoza mentioned previously show, the increase in the crime rate has been a very worrying issue for the whole Mendocinean society. Consequently, having more security and being protected has been the most mentioned reason to justify moving to a gated community within the group of structural reasons for the arrival of gated communities. Seventy per cent of the residents interviewed mentioned security as the reason why they moved to Palmares. This is expressed in the following dialogue (all the interviewees have pseudonyms to protect their privacy):

> *Sonia*: What were the motives that made you decide to move to Palmares?
> *Juan*: I bought the plot because of security …
> *Sonia*: And because of anything else?
> *Juan*: No, because of security … well, I liked it … and I wanted to come far away from the centre … but above all I was already seeing the wave of violence and so I wanted to look for a safe place. (Juan, resident of Palmares)

In this sense, the residents of Palmares do not think about the segregation process that is implicated in the arrival of gated communities, but they see their move to the gated community as a way of having a better life (especially more protected) for them and their families.

On the other hand, insecurity in the 'open city' is such a worrying issue that it has become the prime justification for the existence of gated communities. Many people from the surrounding areas would like to live in a gated community, but cannot afford it. Consequently, they have even thought about the possibility of closing their own neighbourhoods. This is illustrated by the following:

> I hope this neighbourhood also becomes private … It would be so nice! (Cristobal, a neighbour from outside Palmares)

Due to the wave of insecurity, security devices are installed over the whole city making it like a prison. As Marcuse (1993, p. 361) points out "… fences around developments are ubiquitous; whether it be luxury coops or public housing, each cluster wishes to be protected from intrusion from the outside. The scale of the phenomenon exceeds anything heretofore seen". This leads to speculation on the social consequences of such a future, with all the neighbourhoods demarcated by barriers and fences to stop those who are unwanted or look suspicious.

Insecurity and fear of crime have increased in the last four or five years, so the reasons for moving to a gated community before that time are likely to have been different. In this sense, older residents of Palmares mentioned the desire for having more comfort at home (a garden and big house), the search for tranquillity (less noise and traffic), and the importance of buying a house that will have a high market value in the future as other reasons for moving there:

> We lived in a flat with little kids and we wanted a house, with garden and courtyard. (Victor, resident of Palmares)

> We started looking for a more comfortable house ... and we couldn't find anything close [to where we lived] and this project came out ... in the beginning we saw it as something for the weekend ... but while the project was progressing ... we thought 'this is not a neighbourhood for weekends' and ... we built the house to our taste so that when we were about to finish it, we decided to move here permanently and leave the other house where we were living ... (Alvaro, resident of Palmares)

On the other hand, according to the interviewees, living in Palmares gives prestige and status, where status is the social position that gives positive or negative privileges and is expressed through social practices, while prestige refers to the subjective evaluation of somebody's status. Marcuse (1993, pp. 360–361) reinforces this idea saying that the "neighbourhood has become more than a source of security, the base of a supportive network, as it has long been; it has become a source of identity, a definition of who a person is and where she or he belongs in society" and he adds "there is a great deal of congruence between residential location and economic position" (Marcuse, 1993, p. 361).

A resident of Palmares expresses what Marcuse points out about the relation between neighbourhood and identity and social homogeneity through her experience:

> *Sonia*: Which do you think is nowadays the most important reason to live in a gated community?
> *Clara*: Well, what I have heard, but it is not my case, is that it is safer and apart from that to belong to certain groups, identify ... It is not the same saying that you live in Godoy Cruz ... Look, there is something symptomatic: I can say that I live either in Godoy Cruz or in Palmares. If I say that I live in Godoy Cruz, I am lazy and if I say that I live in Palmares I am different, even rich and happy, without problems and this brings a bit of envy and they look at you as someone different. (Clara, resident of Palmares)

When residents were asked whether they thought that status could be a reason for moving to a gated community they replied that it could, and that some people were motivated in this sense. But they always denied that it was their own reason. A resident of Palmares expresses this idea in the following way:

> The benefits of living in a gated community ... the security and some time ago, living in Palmares seemed to give a distinct social status ... but for me ... I don't

care about this ... I am not interested ... the one who lived here [Palmares] was somebody with an important economic situation ... not in my case as I built my house with a very big personal effort ... with my job... (Marisa, resident of Palmares)

Neighbours from the surrounding areas also perceive that gaining status is closely related to the act of moving to Palmares. A resident of one of the closer neighbourhoods expresses this through the following conversation:

Sonia: Do you think that there is a strong division between Palmares and the rest?
Rosa: Yes.
Sonia: In what sense?
Rosa: In the sense that they have status and we don't.
Sonia: What do you mean by that?
Rosa: That they have money and we don't. (Rosa, resident of a lower-middle-class neighbourhood)

However, while status is something that residents of Palmares think they can acquire by moving to the gated community, many in the outside community believe that some people can move there on account of the status they already possess.

Analysing the responses of the interviewees about their reasons for moving to Palmares, the theoretical assumptions are corroborated. Security is the most important motive within the structural causes and getting status and a better life quality appear as the subjective reasons for moving to the gated community. As the structuration theory states there are external facts that influence the actors' decisions; but actors also have some room for action in deciding on their residential location. This opens up the question: to what extent is segregation a voluntary or an involuntary act performed by the actors?

Regarding the two groups who are divided by the walls of the gated community, it is interesting to analyse how these two groups perceive each other and, in the case of the residents of Palmares, the difference in how they describe themselves and how the neighbours from outside describe the former. In the first case, Palmares' residents describe themselves as 'middle class', 'workers', 'middle and upper-middle class', but not rich. However, they speculate that outsiders might think that:

We are corrupted, thieves that steal and build nice houses in Palmares ... (Fernando, resident of Palmares)

Another resident of Palmares also expresses what she thinks the neighbours from outside might think about them in this dialogue:

Sonia: As you go to Fusch neighbourhood [for shopping], what do you think people from there think about people who live in Palmares?
Clara: What the whole world thinks about it ...
Sonia: What is it?
Clara: The rich people, that we are rich ... that we establish these differences or who are looking for a level to which ... I don't know if they want to ... but they cannot accede ... I feel it, at least I feel it... (Clara, resident of Palmares)

On the other hand, the surrounding neighbours, without differences regarding their social class, describe the residents of Palmares as 'rich', 'very rich' or 'selected people with high incomes'.

The borders of the gated community lead to the existence of two different worlds: one within the walls and the other outside them. The knowledge, thus, that both groups have of the other is limited. When the residents of Palmares are asked about the surrounding areas they refer either to the shopping mall or to the slum or to the poor neighbourhood crossing the highway, but hardly ever to the middle- or lower-middle-class neighbourhoods closer to them. Usually they do not know the names of the surrounding neighbourhoods. With regard to the poor neighbourhoods and the slum they usually say: "this is the bad part of the neighbourhood", referring to Palmares, or they say that they are 'ugly'. Moreover, they feel worried about them, as expressed by a resident of Palmares:

> ... on the issue of the marginal neighbourhoods that are surrounding Palmares I think that it is something that in the long term might bring upon us some problems and it will become necessary to look for a solution [to them]. (Victor, resident of Palmares)

Most of the residents of Palmares have a very limited knowledge of the surrounding community because they do not have any kind of contact. The majority of them do not perform any activity in the surrounding neighbourhoods or with people who live there. On the other hand, the people from outside do not have any contact with the gated community and most of them have never been inside the walls. Moreover, it is important to demolish the myth that gated communities provide jobs for the surrounding community as usually the workers commute in from places outside the local neighbourhoods. In addition, people from the slum said that they could not ask for a job in Palmares as they were discriminated against because of the neighbourhood in which they live.

When the residents of Palmares are asked about their address they avoid mentioning Palmares and they prefer to say 'Godoy Cruz' (the municipality) as they think when people know that they live in Palmares they always want to charge them more for goods or services (e.g. car repair). In addition to this, sometimes they do not want to say where they live as they do not want to make social differences explicit, as expressed by a resident of Palmares:

> ... there are people that can take it badly. The thing [social situation] is bad as to go on saying that you live in Palmares. (Victor, resident of Palmares)

On the other hand, there are cases in which the residents of Palmares manifest (implicitly or explicitly) having felt discriminated against when other people know that they live in Palmares. One resident of Palmares relates her experience:

> *Alejandra*: I have been working for 7 years in a [primary] school [as a teacher] and since I moved [to Palmares] the director of my school if I complain or say something [she] tells me I should stay at home and leave the post to others ... I have felt discriminated against in my place of work.
> *Sonia*: And why do you think this happens?
> *Alejandra*: For envy ... I can't see any other reason. (Alejandra, resident of Palmares)

Another boy from Palmares put it like this:

> Later I got used to it, above all because of people's prejudices: 'ah! The guy who lives in Palmares?' but after that you don't care anymore about it. (Fernando, resident of Palmares)

Some aspects about feeling discriminated against came up naturally in the interviews when the interviewees have suffered this, but otherwise, discrimination issues were not mentioned by the interviewees if they were not asked about them. With regard to segregation, the situation was even more implicit in conversation and it was never mentioned naturally. The approach to it had to be handled very discreetly, by asking about increasing social differences.

Most of the time, residents of Palmares do not agree that gated communities might contribute to urban social segregation. Sometimes, they accept that segregation is implicit but they see it as a natural condition of capitalist society or 'the social situation'. One resident of Palmares explains the situation in this way:

> *Juan*: I did not feel discriminated against when I did not live in a gated community by the ones who lived in a gated community ... maybe because I thought that once I would be in a gated community, as maybe one day I would have an SUV (or Land Rover).
> *Sonia*: So you don't think that the gated communities discriminate, do you?
> *Juan*: No, I think they discriminate but because of the social situation that the country is facing nowadays, because if all were satisfied with their jobs and with a minimum salary that allowed them to survive fine, they would not look everywhere, but as there is so much poverty now, so the differences are always marked ... the differences are marked a lot ... (Juan, resident of Palmares)

This kind of situation has an impact on social relations, in the way they are constructed and established. There are two sides of the coin, represented by those who live inside the walls, and those who live outside. They both appear as enemies to each other due to the privileges that one side has and the other lacks. Probably from the director of the primary school's viewpoint Alejandra is perceived as 'a privileged and rich woman' as she lives in Palmares and consequently she would not need to work (not even as a way of satisfying a personal need).

On the other hand, the residents of the surrounding neighbourhoods have also reported feeling segregated by the residents of Palmares in some cases. As one man expresses:

> They do not have—the ones from Palmares—a relationship with the surrounding areas ... they do not want to have it ... I don't know why ... like a neighbour that you have for 10 or 20 years and doesn't talk to you. (Cristobal, resident of a surrounding lower-middle-class neighbourhood)

Neighbours have also mentioned having felt discriminated against, with an extreme case of a man who is going to sue Palmares due to discrimination as he explains:

> *Leopoldo*: ... the security system, which can be very efficient, it is efficient for them [residents of Palmares], but it is terrible for us [the surrounding neighbours] ... we are about to start a legal action about discrimination.
> *Sonia*: Why?
> *Leopoldo*: Why? Do you know the meaning of Kelpers[7]? Well, we are the Kelpers and the British are the ones inside. Because all the controls, everything related to security is managed through this gate [he lives opposite the gate, which is the service entrance], from the gate to outside. So, from the gate to inside all who are authorised can go, the ones who have car insurance ... the others stay outside ... they [residents of Palmares] cannot have people eating in the streets [referring to the construction workers]. I do have to put up with them because I am not on the same level as them, because they have their own private security. The cars that cannot go inside because they do not have their papers in order I have to stand them ...[they are left outside the gate of Palmares]. (Leopoldo, resident of a surrounding middle-class neighbourhood)

In December 2001 (when the then president De la Rua quit and Argentina passed through one of its worst social and political crises) much looting took place in different cities. Residents of gated communities all over the country were afraid of being the target of looters and so the surrounding poor neighbourhoods became an object of charity and in same cases residents of gated communities started to provide them with food or services. This also happened in the case of Palmares and residents started collecting food to provide to the children of the slum near Palmares.[8] Talking about this help to the poor people, a resident of Palmares expresses:

> [We collaborate] with a double meaning ... that if we are fine, we want the ones next to us, the neighbours, to be fine [also] or help them to be a little better and ... [on the other hand] that they do not see us as enemies ... (Samuel, resident of Palmares)

The feeling of being different or even the enemy is always present in the two sides' discourses. Social differences between inside and outside are perceived by all actors involved. Here are the opinions of two women from outside Palmares and a man from Palmares:

> You live there and I live here and don't get involved with me ... if one does not have an invitation she/he cannot go inside [the gated community] but here they can go inside. (Catalina, resident of the slum)

> Differences are noticeable ... it is not the same a neighbourhood house [in this area] and a house in Palmares ... the land prices, construction prices, etc. ... prices are higher there than here ... (Lucia, resident of a surrounding middle-class neighbourhood)

> Nowadays, these social differences are unfortunately much bigger ... there are poor people and rich ... the problem is that now many people are much poorer and the middle-class people dropped to poor-class ... (Alvaro, resident of Palmares)

Conclusions

The existence of social differences is a characteristic of the capitalist society and this is manifested in the city through the existence of different groups who live in different locales, and thus, through a process of urban social segregation. Gated communities have to be considered in the light of these urban segregationist tendencies. However, the arrival of gated communities with their physical barriers make social differences more evident in the city landscape. This has led to a new kind of urban social segregation that is legitimised by law and has different impacts. The case study strongly suggests that the negative consequences are more appreciable and deeper-felt than the positive ones.

The analysis of this case study from an intermediate city in Argentina is relevant as it will widen our understanding of this urban phenomenon in the country, and enable a comparison with Buenos Aires, where gated communities have had an impressive development. Moreover, the link between gated communities and urban social segregation enables the construction of a theoretical framework around the concept of urban social segregation. This paper has suggested considering the ideas of the structuration theory, as structural and subjective forces need to be considered together as both have influenced the growth of gated communities.

This study has provided empirical data to support the theoretical assumption that there is a link between the development of gated communities and urban social segregation. The qualitative research has recorded opinions, feelings and perceptions not only of the residents of the gated community but also of the outside community. It has showed that on the one hand, the residents of the gated communities feel more comfortable and more secure as they have all kinds of security devices, but they also feel envied or discriminated against by the ones outside for having these 'benefits' (chiefly security and comfort). Furthermore, they feel embarrassed knowing that they have many benefits or privileges that others cannot have. On the other hand, neighbours of the surrounding areas feel discriminated against and segregated, as they do not have the same services and benefits as inside. Moreover, they know that they cannot afford to live in a gated community.

For each group there is no knowledge of the other group and a lack of contact. Consequently, social distances become more marked and the feelings against the other group become stronger. Two different worlds appear with almost nothing in common: the ones inside and the ones outside the walls of the gated communities. Hence, there are two sides of the coin. Those who live in gated communities are protected from the outside world, especially from urban violence and crime, while others cannot live in such a residential development as they cannot afford it and, consequently, they suffer from insecurity and fear of crime in their everyday lives. The analysis of the opinions, feelings and perceptions of the two groups has showed that there is a link between urban social segregation and gated communities in the sense that gated communities contribute to the development or reinforcement of the process of social segregation that characterises the capitalist city.

The question is who segregates whom? Do those who live in gated communities segregate themselves or have they ended up being segregated by the rest of society in the process of securing themselves a better lifestyle? The answer to 'who segregates whom' can take two different forms according to the viewpoint of the speaker. No matter who is segregated and who is the segregator the important issue to consider is that there is a division of the society and social divisions have consequences on the development of a prosperous society, and that gated communities do contribute to exacerbate the process of urban social segregation. Is it too late to try to change this pattern?

Acknowledgements

This paper is based on the research conducted during 2003 for a PhD thesis 'Social practices in gated communities: urban social segregation as a myth or a reality? A case study from Mendoza, Argentina', that the author is currently writing at the Development Planning Unit, University College London.

The author would like to particularly thank all the interviewees for their help and acceptance to be interviewed, for giving their opinions about the topic and sharing their experiences. Thanks are also due for the interesting comments on a previous version of this paper made by anonymous referees. Finally, thanks are due to Matt Chesterton for his help.

Notes

[1] See data of crime rate by province and total of the country in http://www.economia.mendoza.gov.ar/sitios/deie/banco%20de%20datos/social/segpubli/archivos/comseg01.xls (accessed 28 March 2004).

[2] See data of Mendoza from http://www.economia.mendoza.gov.ar/sitios/deie/banco%20de%20datos/social/vivienda/archivos/viv01.xls (accessed 28 March 2004).

[3] See data from the National Institute of Statistics and Censuses (INDEC) www.indec.gov.ar (accessed 28 March 2004).

[4] By autonomy Castells means "the average margin of individual autonomy/social heteronomy as constructed by society; ... to social conditions that represent the social norm, in contrast with people's inability to organise their own lives even under the constraints of social structure, because of their lack of access to resources that social structure mandates as necessary to construct their limited autonomy..." (1998, footnote 9, chapter 2, p. 73).

[5] The poverty line is constructed considering the costs of the basic basket of goods, services and expenses in education, health, clothes and transport for a household.

[6] In the second stage of Palmares the differences among the houses are more visible than in the first one where there is a sort of homogeneity mainly regarding sizes, while in the second stage the houses of the 'nouveau riche', many of whom have not moved yet as the houses are not finished, are really impressive. Many of the latter can be seen from the highway. Most of the families have 1 to 2 plots, but the 'new rich' have 4, 6 or 8 plots.

[7] Kelpers are the people who live in the Malvinas Islands (Falklands).

[8] The slum is called Urundel. The local church is running a 'comedor' (place to eat) for about 100 children of the neighbourhood to go there on Saturdays and have lunch. Some of the residents of Palmares collaborate with food or money. A group of women from Palmares also collected toys for the children of the slum to give to them at Christmas.

References

Arizaga, M. C. (2000) Murallas y barrios cerrados. La morfología espacial del ajuste en Buenos Aires, *Nueva Sociedad*, 166, marzo–abril, pp. 22–32.

Blakely, E. J. & Snyder, M. G. (1997) *Fortress America. Gated Communities in the United States* (Washington DC and Cambridge, MA: Brookings Institution Press and Lincoln Institute of Land Policy).

Blauw, W. (1991) Conclusion, in: E. D. Huttman (Ed.) *Urban Housing Segregation of Minorities in Western Europe and the United States*, pp. 391–402 (Durham and London: Duke University Press).

Borsdorf, A. (2002) Barrios cerrados en Santiago de Chile, Quito y Lima: tendencias de la segregación socio-espacial en capitales andinas, in: L. F. Cabrales Barajas (Ed.) *Latinoamérica: países abiertos, ciudades cerradas*, pp. 581–610 (Guadalajara, México: Universidad de Guadalajara, UNESCO).

Bragos, O., Mateos, A. & Pontoni, S. (2002) Nuevos desarrollos residenciales y procesos de segregación socio-espacial en la expansión oeste de Rosario, in: L. F. Cabrales Barajas (Ed.) *Latinoamérica: países abiertos, ciudades cerradas*, pp. 441–480 (Guadalajara, México: Universidad de Guadalajara, UNESCO).

Cabrales Barajas, L. F. & Canosa Zamora, E. (2002) Nuevas formas y viejos valores: urbanizaciones cerradas de lujo en Guadalajara, in: L. F. Cabrales Barajas (Ed.) *Latinoamérica: países abiertos, ciudades cerradas*, pp. 93–116 (Guadalajara, México: Universidad de Guadalajara, UNESCO).

Caldeira, T. P. d. R. (2000) *City of Walls. Crime, Segregation and Citizenship in Sao Paulo* (Berkeley, CA: University of California Press).

Carvalho, M., Varkki George, R. & Anthony, K. (1997) Residential satisfaction in *Condominios Exclusivos* (Gate-guarded Neighborhoods) in Brazil, *Environment and Behavior*, 29, pp. 734–768.

Castells, M. (1998) *End of Millennium. The Information Age: Economy, Society and Culture* (Oxford: Blackwell).

Ciccolella, P. (1999) Globalización y dualización en la Región Metropolitana de Buenos Aires. Grandes inversiones y restructuración socioterritorial en los noventa, *EURE*, 25(76), pp. 5–27.

DEIE (Dirección de Estadísticas e Investigaciones Económicas) Gobierno de Mendoza, Argentina. Available at http://www.economia.mendoza.gov.ar/sitios/deie.

INDEC (Instituto Nacional de Estadísticas y Censos) Ministerio de Economía, República Argentina. Available at http://www.indec.mecon.gov.ar.

Giddens, A. (1984) *The Constitution of Society. Outline of the Theory of Structuration* (London: Polity Press).

Greenstein, R., Sabatini, F. & Smolka et al. (2000) Urban spatial segregation: forces, consequences, and policy responses, *Land Lines*, 12(6), (MA: Lincoln Institute of Land Policy).

Ickx, W. (2002) Los fraccionamientos cerrados en la Zona Metropolitana de Guadalajara, in: L. F. Cabrales Barajas (Ed.) *Latinoamérica: países abiertos, ciudades cerradas*, pp. 117–141 (Guadalajara, México: Universidad de Guadalajara, UNESCO).

Janoschka, M. (2002) Urbanizaciones privadas en Buenos Aires: ¿hacia un nuevo modelo de ciudad latinoamericana?, in: L. F. Cabrales Barajas (Ed.) *Latinoamérica: países abiertos, ciudades cerradas*, pp. 287–318 (Guadalajara, México: Universidad de Guadalajara, UNESCO).

Landman, K. (2002) Gated communities in South Africa: building bridges or barriers? Paper presented at the International Conference on Private Urban Governance, Mainz, Germany, 6–9 June.

Low, S. M. (2000) *The Edge and the Center: Gated Communities and the Discourse of Urban Fear*. Available at http://062.cpla.cf.ac.uk/wbimages/gci/setha1.html.

Low, S. (2003) *Behind the Gates* (New York and London: Routledge).

Marcuse, P. (1993) What's so new about divided cities?, *International Journal of Urban and Regional Research*, 17, pp. 355–365.

Marcuse, P. (1994) Walls a metaphor and reality, in: S. Dunn (Ed.) *Managing Divided Cities*, pp. 41–52 (Keele: Ryburn Publishing).

Marcuse, P. (2001) Enclaves yes, ghettos, no: segregation and the state. Paper presented in the International Seminar on Segregation in the City Lincoln Institute of Land Policy, Cambridge, MA, 25–28 July.

Massey, D. & Denton, N. (1993) *American Apartheid: Segregation and the Making of the Underclass* (Cambridge, MA: Harvard University Press).

Mingione, E. (Ed.) (1996) *Urban Poverty and the Underclass: A Reader* (Oxford: Blackwell).

Prévot Schapira, M.-F. (2000) Métropoles d'Amérique Latine: de l'espace public aux espaces privés, *Cahiers Des Amériques Latines*, 35, pp. 15–19.

Rovira Pinto, A. (2002) Los barrios cerrados de Santiago de Chile: en busca de la seguridad y la privacidad perdidas, in: L. F. Cabrales Barajas (Ed.) *Latinoamérica: países abiertos, ciudades cerradas*, pp. 351–369 (Guadalajara, México: Universidad de Guadalajara, UNESCO).

Svampa, M. (2001) *Los que ganaron. La vida en los countries y barrios privados* (Buenos Aires: Biblos).

Thuillier, G. (2000) Les quartiers enclos á Buenos Aires: quand la ville devient country, *Cahiers Des Amériques Latines*, 35, pp. 41–56.

Wilson, W. J. (1987) *The Truly Disadvantaged; the Inner City, the Underclass and Public Policy* (Chicago: The University Press of Chicago).

Wilson-Doenges, G. (2000) An exploration of sense of community and fear of crime in gated communities, *Environment and Behavior*, 32, pp. 597–611.

Gated Communities: Sprawl and Social Segregation in Southern California

RENAUD LE GOIX
Department of Geography, University Paris 1 Panthéon-Sorbonne, Paris, France

(Received October 2003; revised April 2004)

KEY WORDS: Gated communities, urban sprawl, segregation

Introduction

Gated communities, which are walled and gated residential neighbourhoods, have become a common feature in US metropolitan areas. Based on an empirical study in the Los Angeles region, this paper focuses on how gated communities, as a private means of provision of public infrastructure, produce increased segregation at the local scale. It aims to trace the ways local governments usually favour the development of this form of land use to pay for the cost of urban sprawl, while indeed producing social diseconomies for the whole metropolitan area.

The social sciences literature about gated communities has been highly publicised, and three types of arguments are now part of a general theoretical discourse, which especially focuses on the relationship between gated communities and social segregation. First, gated enclaves are described both as a physical and obvious expression of the post-industrial societal changes (fragmentation, individualism, rise of communities), as part of a commoditisation trend of urban public space (Dear & Flusty, 1998; Sorkin, 1992), and as a penetration of ideologies of fear and security supported by economic and political actors (Davis, 1990, 1998; Flusty, 1994; Marcuse, 1997). A second set of arguments presents gated communities as symptoms of urban pathologies, among them social exclusion is considered to be pre-eminent. Voluntary gating and the decline of public spaces in cities are seen as being detrimental to the poorest social classes (Blakely & Snyder, 1997; Caldeira, 2000; Glasze *et al.*, 2002). Finally, the rise of private enclaves is argued to be a 'secession' by an elite opposed to the welfare redistribution system (Donzelot, 1999; Donzelot & Mongin, 1999; Jaillet, 1999; Reich, 1991), given the assumption that public provision of services is inefficient (Foldvary, 1994). The debate about gated enclaves and segregation has been lively despite a lack of empirical arguments to sustain it, as it is difficult to gather a representative sample of gated communities at a local scale.

This research derives from the above outline of arguments. It seeks to provide some evidence of the impact of these communities on segregation patterns within the metropolitan region of Los Angeles. To introduce how gated communities produce social exclusion, it is of interest to recall how developers usually design them as homogeneous social environments. The appeal of gated communities is inspired by the historical private estates found near industrial-era cities, such as Llewellyn Park near New York, associated with an anti-urban ideal (Castells, 1983; Jackson, 1985). Today, gated enclaves are mostly commoditised suburban neighbourhoods for the upper and middle class, emphasising a 'community lifestyle' (Blakely & Snyder, 1997). Their promotion typically focuses on sport and leisure amenities and family life. Furthermore, they are Common Interest Developments (CIDs), aiming to protect property values through design policies and Covenants, Conditions and Restrictions (CC&Rs). Together with landscaping and architectural requirements, subjective criteria of social preference are common in many CIDs (Fox-Gotham, 2000; Kennedy, 1995; McKenzie, 1994; Webster, 2002), thus helping to maintain a homogeneous social pattern. Furthermore, CIDs are public actors because of the nature of their provision of a public service to the residents and their right to collect a regular assessment. At the same time they act as private governments, based on a private contract (CC&Rs) enforced to protect property values (Kennedy, 1995; McKenzie, 1994).

But gated communities are far more than a regular CID. Excluding themselves from the public realm, gated communities are then referred to as a 'club' (Webster, 2002). For the residents, all being members of the 'club', gating a neighbourhood can be conceived in a first instance as a pre-emptive attempt to protect the neighbourhood. Residents are supplied with their own security, roads, amenities, etc., in a private governance effort to avoid the spillovers of urban residential and industrial developments: crime, increasing through traffic, free-riding of the amenities, urban decay and decreasing property values due to unwanted land-use.

This paper proposes to analyse this pre-emptive protection of the neighbourhood as being detrimental to the neighbours of a gated community and the adjacent urban communities. This proposition requires considering the broader theoretical context of the production of urban space in a capitalist city, and the genesis of the urbanisation process within the capitalist mode of production. This can be described as a land-use system consisting in interpenetrating private and public spaces governed by complex patterns of property rights. These spaces are the outcomes of location strategies of actors considering "dense polarised differential locational advantages through which the broad social and property relations of capitalism are intermediated" (Scott, 1980). The capitalist production of urban space by private firms and homeowners, making individually optimal decisions, has a social cost and generates spillover effects, such as pollution, sprawl, congestion, competition for land uses, land speculation, free-riding. Interpreted as a market failure (Bator, 1958), such externalities represent a cost for the society as a whole. Following this theoretical thread, gating a neighbourhood can be conceived in a first instance as a private pre-emptive solution of market failures. It supplies the residents with their own private governance effort to avoid the spillovers of urban residential and industrial developments. On the other hand, gated communities also produce spillover effects on their neighbours, which this paper aims to address with a special interest in the impact on social patterns.

In this context, it is assumed that the specificity of gated communities does not derive from property-owners' association status, a now dominant form of housing in the US (McKenzie, 1994). Indeed, it relies on the physical border, which interacts with the territorial nature of the urban space. On the one hand, access control and security features represent a substantial cost for the homeowner, not only for the cost of building the infrastructures, but also for their maintenance. On the other hand, the private access acts as a guarantee of the exclusive use of a site, which favours site rent and property value and creates a desirable place to live in (Le Goix, 2002). As a consequence, the question does not address the CID that lies behind the gate, but the effects of gating. Gating can then be analysed as a border between several territorial systems: the systems of the city and adjacent neighbourhoods versus the system of the gated enclave. This paper analyses the sprawl of gated communities in southern California, and evaluates its social impact. It is based on a methodology to assess the impact of gating over the social and ethnic patterns of residents, both inside the gated enclave and beyond the walls.

In the first part of this paper, a comprehensive study of the diffusion of gated communities in southern California leads to insights about the connections between gating and urban sprawl, since public governments tend to transfer the costs of urbanisation to private developers. Socio-economic and ethnic spillover effects are analysed in the second part. The demonstration of increased segregation associated with gating relies on a dissimilarity index and discontinuities' mapping, which relevant methodology will be hereafter explained.

Gated Communities and Urban Sprawl

The diversity of gated communities has to be taken into account in order to assess the extent of the market accurately. Blakely & Snyder (1997) have identified three major types of gated communities: elite or golden-ghetto communities based on prestige, lifestyle communities where gates assure the exclusive access to leisure facilities, and 'security zone communities' where safety is the main concern of residents and now include several low-end neighbourhoods retrofitted with gates to promote their safety and control gang activities.

The Location of Gated Communities

Because of the lack of a comprehensive survey of gated communities on a local scale, this research is based on a database derived from the same sources as a prospective homebuyer would use. Once integrated within a Geographical Information System (GIS) with 2000

Census data, the diversity of the market can be assessed, as well as the location of gated communities, their social patterns and their impact. Accompanied by field surveys, interviews with gated communities and local officials, the most important sources for locating gated neighbourhoods were *Thomas Guides®* maps plotting gates and private roads, real-estate advertisements in the press and in real-estate guides, and County Assessor's maps. Thus 219 gated communities built before 2000 have been identified in seven counties (Los Angeles, Riverside, Orange, Ventura, San Bernardino, Santa Barbara and San Diego).

Using the latest results from the *2001 American Housing Survey*, Sanchez *et al.* (2003) accurately estimate that 11.7 per cent of the households are in walled, fenced and access-controlled communities in the Los Angeles Metropolitan Area, based on a national sampling of households. It is relevant to mention here that the research presented herein relies on a more restrictive sampling of gated communities, designed to exclude the condominiums and secured apartment complexes, as they do not include privatised public spaces, according to Blakely & Snyder's definition of gated communities (1997). Usually in vertical co-ops and condominiums the common areas are limited to a parking lot, a shared garden or a swimming pool, which do not meet the definition of public spaces (streets, places, sidewalks, parks, beaches).

For the sample of gated communities for which the size is accurately known (Figure 1), the number of dwelling units located behind gates in 2000 can be estimated to be 80 000 (an estimate of 230 000 inhabitants), or 1.5 per cent of the housing stock, and increasing at a fast pace. In 2001, according to the *New Home Buyers Guides*, this market represented a 12 per cent average of the new homes' market in southern California, but 21 per cent in Orange County, 31 per cent in San Fernando Valley and 50 per cent in the desert resort areas of Palm Springs.

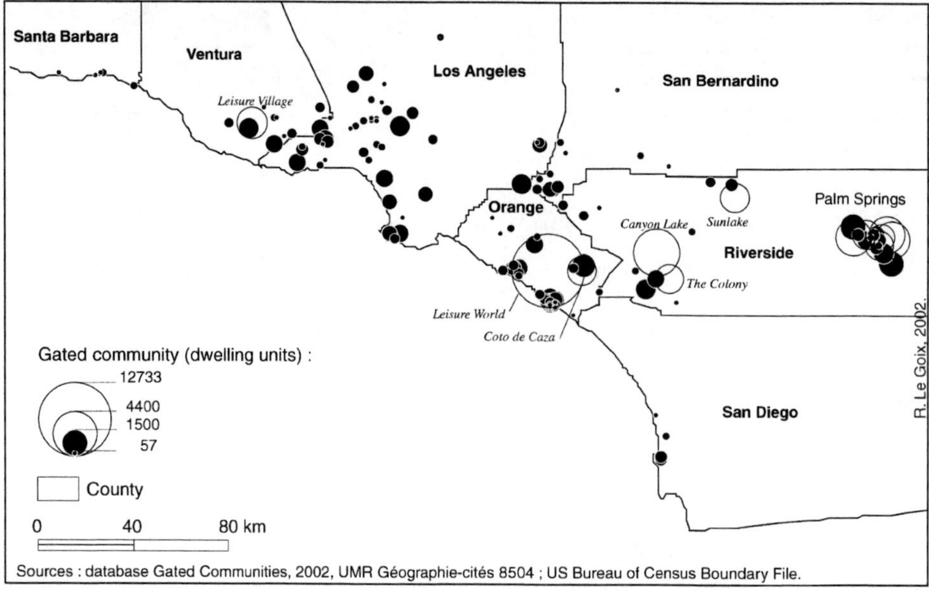

Figure 1. The size and location of gated communities in southern California. *Source*: Database Gated Communities, 2002, UMR Géograpie-cités 8504; US Bureau of Census Boundary File.

Three factors explain the location of gated communities. First, their locations tend to maximise location rents. Ocean fronts in Santa Barbara, Malibu, Newport Beach or Dana Point in Orange County, secluded hill areas in Palos Verdes peninsula or in Malibu Mountains, and finally desert and resort sites (Palm Springs) are the favourite locations.

Second, as most of the leisure-oriented residential developments use a large amount of space, the availability of land is an important factor. Leisure facilities and amenities indeed consume a lot of space, and both developers and residents favour large individual lots clustered in a surrounding setting of large open space. Every large gated community such as Leisure World (19 000 inhabitants and 6 clubhouses) and Canyon Lake (9000 inhabitants), as well as seven gated neighbourhoods of more than 1500 housing units in Palm Springs, are lifestyle communities or retirement communities located in remote settings favouring scenic views and the intimacy of residents. The secluded and oasis locations serve the same goal as the gate towards isolation from the urban social context.

Smaller gated communities are clustered near the central places of the urban region, in the north of Los Angeles County (west Los Angeles, Burbank and San Fernando Valley), in Irvine and Anaheim in Orange County or in the western side of San Bernardino County (Chino and Ontario). Some of them are former open neighbourhoods, which have opted for the gating, like Fremont Place or Brentwood Circle on Sunset Boulevard (Moore, 1995a, 1995b). But small gated communities are often in-fill developments of vacant land in older urbanised areas, such as the upper-scale community of Manhattan Village (520 housing units) in Manhattan Beach, or the new middle-class communities named Stonegate (57 housing units) and Lori Lane (40 housing units) by Kaufman & Broad in Anaheim and Garden Grove.

Finally, location is driven by the social environment. It is assumed that gated communities are tailored to fit to specific prospective buyers and located within a consistent social environment. A former study showed that gated communities are located within every kind of middle-class and upper-class neighbourhood, and are now available for every market segment (Le Goix, 2002). Half of them are located within the rich, upper-scale and mostly white neighbourhoods, and one-third are located within the middle class, average income and white suburban neighbourhoods. As evidence of the social diffusion of the phenomenon, 20 per cent of the gated communities surveyed are located within average and lower income Asian or Hispanic neighbourhoods, especially in the northern part of Orange County and in the north of San Fernando Valley.

A Diffusion of Gated Communities According to Suburban Sprawl Patterns

The spatial distribution of gated communities is linked with the urban sprawl. A chronological cartography (Figure 2) of four different stages shows evidence of a diffusion process within the Los Angeles region. Each map describes the situation at a date when important change occurred in the Los Angeles development. The first gated neighbourhoods were developed in 1935 in Rolling Hills and in 1938 in Bradbury, and some well-known gated communities were built early after the Second World War, like the upper-scale Hidden Hills (1950), and the original Leisure World at Seal Beach (1946) housing veterans and retired people in Orange County. Before 1960, about 1700 housing units were gated in the Los Angeles area, increasing up to 19 900 in 1970 because of the developments of major gated enclaves like Leisure World (1965) and Canyon Lake (1968).

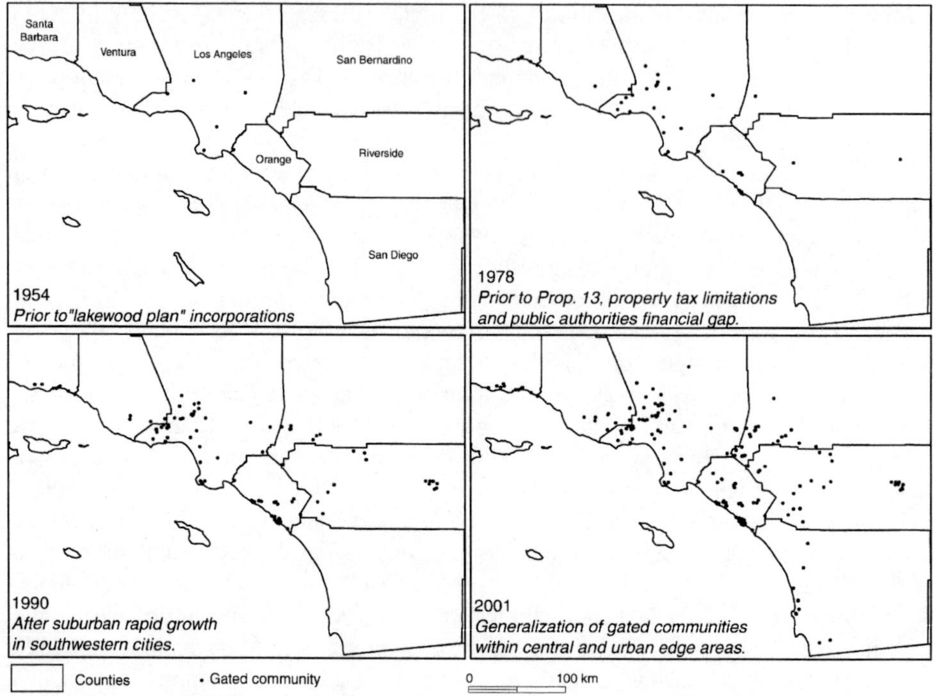

Figure 2. The diffusion of gated communities in southern California.

After 1970, the new developments were smaller than they used to be in the 1960s, and the growth rate decreased: 31 000 gated units in 1980, 53 000 in 1990 and 80 000 in 2000.

The diffusion pattern of residential homogeneous suburban communities is related to the suburban growth, an anti-fiscal posture, and the municipal fragmentation dynamic that have affected the Los Angeles area since the 1950s. In Los Angeles, this trend has been motivated by an anti-fiscal posture, by the means of municipal incorporations as the Lakewood residential development first experienced when it became an autonomous city in 1954. Many municipal incorporations were designed to avoid paying the costly county property taxes while charged a lower assessment by the city and getting a better control over local development (Miller, 1981). A second diffusion step came with the 1978 'taxpayers' revolt', when homeowners became the driving force for a property taxes roll back known as Proposition 13 (Purcell, 1997). Meanwhile, the tax limitation was increasing the need for public governments to attract new residential developments and wealthy taxpayers under their jurisdictions, thus supporting gated communities as perfect 'cash cows' (McKenzie, 1994). A third spatial diffusion pattern of gated enclaves is connected with the trend of rapid growth in southern California, sustained by massive population flows driven to the Sun Belt cities during the 1980s (Frey, 1993).

According to Figure 2, three diffusion processes of gated communities have occurred in the area:

- A diffusion by contact between zones where gated communities were previously developed. Hidden Hills in the western part of LA County, Eldorado (1957) in

Palm Springs and Indian Wells, or Niguel Shore (1975) in Dana Point played a key role as local landmarks, soon surrounded by other gated enclaves imitating them;

- Preferred locations are where site rental is maximised, explaining the multiplication of lifestyle communities favouring seaside locations (Santa Barbara County, Dana Point, Newport Beach, etc.);
- A diffusion outlining the polycentric pattern of edge-cities, with clusters of gated communities near areas like South Orange County and its dynamic high-tech economy in Irvine, as well as in the San Fernando Valley and Burbank. These dynamic technopoles provide a massive flow of potential buyers. Large urban private developments such as Irvine being designed as innovative privately operated communities, although supported by public authorities (Forsyth, 2002; Garreau, 1991), it is not surprising to find gated communities in such an environment of mixed governance.

A Diffusion Based on a Public-Private Partnership

Most gated communities were built within unincorporated areas, but some have since incorporated as their own municipalities like Bradbury and Rolling Hills in 1957, Canyon Lake in 1991, Leisure World in 1999 (Le Goix, 2001), or as a part of a new city. For example Dana Point incorporated in 1989, Calabasas in 1991, where a substantial part of single-family housing developments is gated. Although the municipality acts as an extension of the Property Owners' Associations, the arguments for the incorporation pointed out the desire to control the local land development, and to challenge the trend of the County Board of Supervisors to support new residential subdivisions. Calabasas' incorporation in 1991 is representative of such issues, as the new developments are all gated. Gated communities in the Calabasas Park subdivision challenged in 1987 an extension of 2000 units on unincorporated land, while pushing for the incorporation. As the incorporation had previously failed several times, homeowners became suspicious that the County Board of Supervisors might try to push for new developments, while the Local Authority Formation Commission (LAFCO) was slowing down the incorporation. The incorporation process went back and forth over 11 years, while the County had already approved 4500 new units (Kazmin, 1991a, 1991b; Pool, 1987a, 1987b, 1987c).

Calabasas is a good example of the ambiguous relationships between the public authorities' interests for developing gated communities, and the private homeowners' willingness to live in a secluded and controlled place. Gated communities basically are Planned Unit Developments (PUD), implying that the developer substitutes the public government in planning and building roads, access and utilities lines (Knox & Knox, 1997). As stated in the California *Subdivision Map Act* (Sections 66410 et seq.), the public authority has jurisdiction to regulate and control the development of the project in a subdivision. Once the tentative maps are accepted and the subdivision authorised, the builder replaces the public authority. In the case of Master Planned Communities, such a substitution is comparable to a private provision of public services (McKenzie, 1994), as the developer is required to finance the infrastructures, landscaping and improvements to ensure the consistency of the development with any applicable general plan (Curtin, 2000, p. 83). As a consequence, the overall cost of urbanisation is transferred to the private developer, who consequently makes the final buyer pay for those infrastructures when ultimately purchasing his property. Other tools are also available to transfer the urbanisation costs to the final homeowner, instead of the general taxpayer having to pay for them. Such tools encompass the 'developer's fees' paid by a developer to the public authority to cover the public services improvements needed by every additional unit. Alternatively, the developer may be required to set aside a certain amount of free land, which can be used to build a school or a library once ceded to the public authority. Last but not least, a Community Facility District (a costly special assessment on every property located within their boundaries) can be created to transfer the cost to the local homeowner instead of charging the general taxpayers (Brown, 1991).

These transfers of urbanisation costs to the homeowner are outlining the interest of gated communities in the urban planning process. Because of the gate, no public money can be spent within the gates, otherwise the public access to any public-owned facility located inside the community would be granted and the gates would eventually become useless and fail to achieve their goal. Such issues are documented by the 1992 decision of Hidden Hills to build its city hall outside its gates in order to allow public access (Ciotti, 1992; Stark, 1998). The 1994 *Citizens Against Gated Enclaves (CAGE) vs. Whitley Heights Civic Association* case banned the gating of public streets (Kennedy, 1995). In 1999, Coto de Caza rejected a project to build a public school within its gates because it would have allowed the public inside the gated community (N'Guyen, 1999). As a consequence, no public money can be spent for the maintenance of the private roads since they are gated.

Indeed, the development of gated communities is the result of a market demand for security features fitting a standardised leading offer from the homebuilding industry, but also emerges from a partnership between local governments and private land developers. Both agree to charge the final consumer (i.e. the homebuyer) with the overall cost of urban sprawl, since he will have to pay for the construction and the maintenance of urban infrastructures located within the gates. As compensation, the homebuyer is granted with a private and exclusive access to sites and former public spaces (for example, the lake in Canyon Lake, which is originally a public property leased to the association). Such exclusivity favours the location rent, and can positively affect the property value. On the other hand, it provides the public authorities with wealthy taxpayers, thus considering gated communities as property tax 'cash cows' (McKenzie, 1994).

Assessing the Impact of Gated Communities on Social Segregation

Given the ambiguous relationships between public authorities and gated communities favouring the sprawl of a peculiar form of urbanism, the question is then: how does this affect segregation patterns? As previously mentioned, the private governance and the implementation of restrictive covenants lead to an implicit selection of the owners, through design guidelines, age restrictions or a selective club membership, in order to ensure the homogeneity of the neighbourhood. Access control features reinforce this construction of exclusion, as one can be only from the inside, or from the outside. The following hypothesis can thus be formulated: the gating and the exclusiveness create a border. The border separates two spatial systems: the territorial system of the gated community, and the urban space where it is located. It is assumed that the act of gating is worth its cost, and that it has an effect over the social patterns, the property values, etc., thus making gated communities a desirable residential environment to live in.

The Impacts of Gating

Accordingly, gated communities should differentiate from their immediate vicinity, from which homeowners are trying to protect themselves against negative spillovers (crime, property value decay, etc.), hiding behind gates. However, because the erection of a border implies a two-way relationship between the two adjacent territories, gated communities also produce externalities over their neighbours. Such issues have already been discussed with regard to crime and property value patterns. As the motivations for living in a gated community are mostly driven by the fear of crime and fear of differences (Low, 2001), scholars have studied the impact of gating, although limited by the lack of empirical data. For instance, Helsley & Strange (1999) theoretically demonstrated that gating leads to a relocation of crime outside the gates and within adjacent non-gated communities. Studies were conducted assessing the effect of gating over property values. They demonstrated the protection of gated property values (Bible & Hsieh, 2001; Lacour-Little & Malpezzi, 2001), and the deterrent effect on property values in adjacent communities (Le Goix, 2003).

Herein lies the most well-known effect of gating: its negative impact on property values in non-gated adjacent neighbourhoods, and the theoretical crime redistributions. Such diseconomies may lead to a preventive proliferation of gating in the neighbourhood, and former non-gated communities may have to retrofit with gates if they wish to maintain their property values, and avoid crime redistributions, thus explaining the clustered diffusion pattern previously exposed.

As well as crime and property values, it appears necessary to also address the social externalities of gating. This can be achieved to a certain extent by measuring how homogeneous gated communities can be compared to their neighbours, and on which criteria they differentiate from their vicinity. The paper proposes a method to evaluate the level of socio-economic differentiation occurring where gated communities are present. This is used in order to estimate the effect of gating over social segregation, relying on the following assumption: if the overall differentiations occurring between gated enclaves and their vicinities are higher than the differentiations usually observed in the urban area between two adjacent neighbourhoods, then there is a high probability that gated communities indeed produce increased segregation.

The Discontinuity as a Geographical Concept

For that purpose, the discrepancy between a gated community and its vicinity is defined as a 'discontinuity', in order to focus on whether a higher degree of social differentiation occurs where gates and fences are erected. In its broader definition, a discontinuity is what separates two adjacent spatial systems (Brunet, 1965). Close to the notions of 'border' and 'barriers', the notion of discontinuity was used to study the differentiation processes produced by national borders on demographic patterns (Decroly & Grasland, 1992; Grasland, 1997). Discontinuity is a useful concept because not only does it address the ideas of separation (as a barrier) and segregation, but it also allows the description of urban spaces in terms of differentiation processes produced by, or producing, physical barriers (François, 1995).

Methodological Concerns

However, the effects of gating on segregation patterns are difficult to evaluate for three reasons. First, it is assumed that social patterns in gated communities are almost consistent with their neighbourhood, in order to ensure the attractiveness of the development for potential buyers. As a consequence the method seeks to sort out the effects of walls and gates on each social characteristic (age, race, economic status). It relies on a multivariate analysis and clustering to test whether a gated community boundary fits any sensitive shift in the statistical definition of social areas.

Second, the implementation of this test relies on a function of the adjacency between census areas. As not properly addressed by the classical segregation or concentration indices (Apparicio, 2000), it has been necessary to use a 'dissimilarity index' (Decroly & Grasland, 1992; François, 1995). The dissimilarity index equals the difference between the two contiguous areas i and j on a continuous factor X. The factor X is extracted from a factor analysis, and describes the relative position of each area on a factorial axis produced by the joint effect of independent variables (Principal Component Analysis). A discontinuity appears where a significant level of dissimilarity between two contiguous census areas occurs. It may then be mapped as a segment materialising the level of discontinuity, and compared with the layout of gated community boundaries.

Third, no direct answer can be provided about the level of discontinuity. The analysis is processed at the census block group level, the smallest geographic level where 100 per cent sample data were available. But the shape of census geographical definition does not systematically match the boundaries of gated communities (Le Goix, 2002). This is a severe limitation in spatial analysis since only 30 gated communities exactly fit one or more census block groups (case A). The location of a gated community within a larger block group is a common case (case B). It is also possible to find several gated communities together with other regular neighbourhoods within a single block group. In order to proceed despite this limitation, three different kinds of vicinity levels were defined, given that a segment is a line materialising the topological contact between two block groups (Figure 3):

- A first vicinity level applies to the segments between a gated community and its immediate surrounding, where block group and gated community definitions perfectly fit each other (case A). In this case, this vicinity level will account for the discontinuities associated with every large gated community (basically, every large gated community housing more than 1500 inhabitants).
- A second vicinity level characterises the dissimilarities observed in the environment where a gated community is located. It is in fact interesting to test whether the stronger discontinuity is produced by a gated community or by its surrounding neighbourhoods. The latter means that a gated community might be surrounded by a very homogeneous social or ethnic buffer zone. This level is defined both by the segments in contact with block groups adjacent to the precedent level (case A), and by the segments of any block group where a small gated community is located (case B).

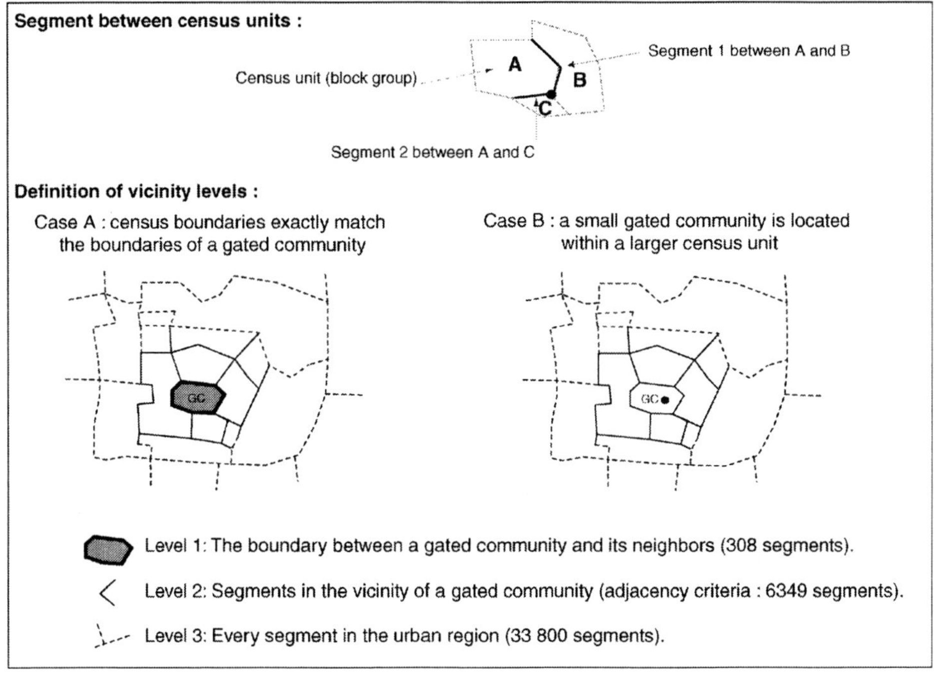

Figure 3. Three vicinity levels around gated communities.

- Finally, in order to set a comparative framework, a third vicinity level comprehends every segment observed between each 12 549 block groups of the area covered by the database. The evaluation of the segregation level at the local scale can then be analysed, everything being equal compared to dissimilarities observed in the whole metropolitan area.

Assessing the Level of Discontinuity

Three main characteristics of the socio-economic differentiation are analysed, using the following variables for each block group of the Los Angeles region:

- Socio-economic status: median property value 2000; owner-occupied housing units (percentage of housing units 2000),
- Ethnicity: white persons; black persons; Hispanic origins; Asian origins; Native American origins (percentage of population 2000),
- Age: less than 18 years old; 18–24 years; 25–44 years; 45–64 years; more than 65 years (percentage of population 2000).

The factorial analysis demonstrates the high level of social separation in the Los Angeles area, with 2000 census data. The first axis (explaining 33.5 per cent of the total variability) describes the ethno-linguistic oppositions and the related effects of age and status: white, aged and wealthy neighbourhoods are opposed to the young, Hispanic and modest neighbourhoods (Figure 4). A second factor (13 per cent) isolates the effect of life cycles and status, and thus opposes neighbourhoods with young 25–44 year-old residents to owner

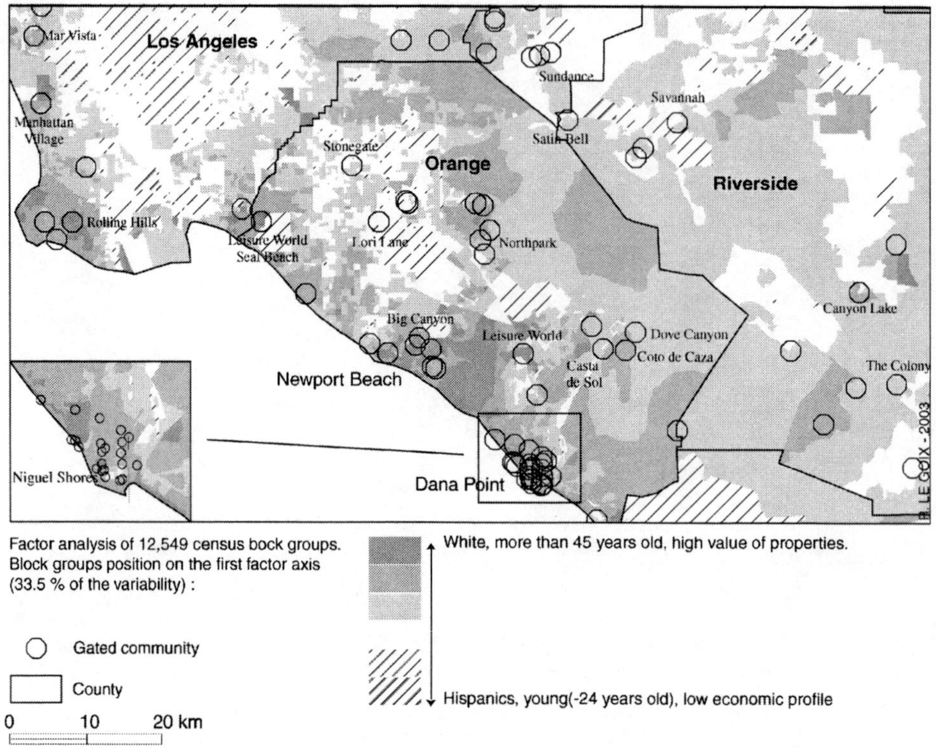

Figure 4. Gated communities and the first socio-economic factor of segregation (south Los Angeles, Orange County and west Riverside County). *Note*: Because of the size of the original maps and edition constraints, only part of the maps are published here.

occupied and families with young children neighbourhoods. A third factor (10.9 per cent) describes the sole effect of ethnic segregation, while it opposes black and Asian neighbourhoods to the mostly white areas, everything being equal regarding the social status. A fourth dimension (9.7 per cent) describes the differences explained by the age, everything being equal regarding the status. Young and middle-aged areas are opposed to very homogeneous neighbourhoods, inhabited by senior citizens, or by high concentration of 18–24 year-olds (campus and military housing), indeed lying behind another type of gate.

A factorial axis being a continuous scale, it allows a comparison of the relative position of block groups. For each segment between two adjacent block groups, the difference between the co-ordinates on each axis was calculated. Considering the absolute value of the dissimilarity indices, the higher the absolute value is, the stronger the discontinuity is.

The Local Increase of Segregation

The map (Figure 5) provides qualitative information regarding the shape of the discontinuities, under the assumption that a continuously shaped discontinuity outlines an independent territorial system, whereas a poorly shaped discontinuity would only outline

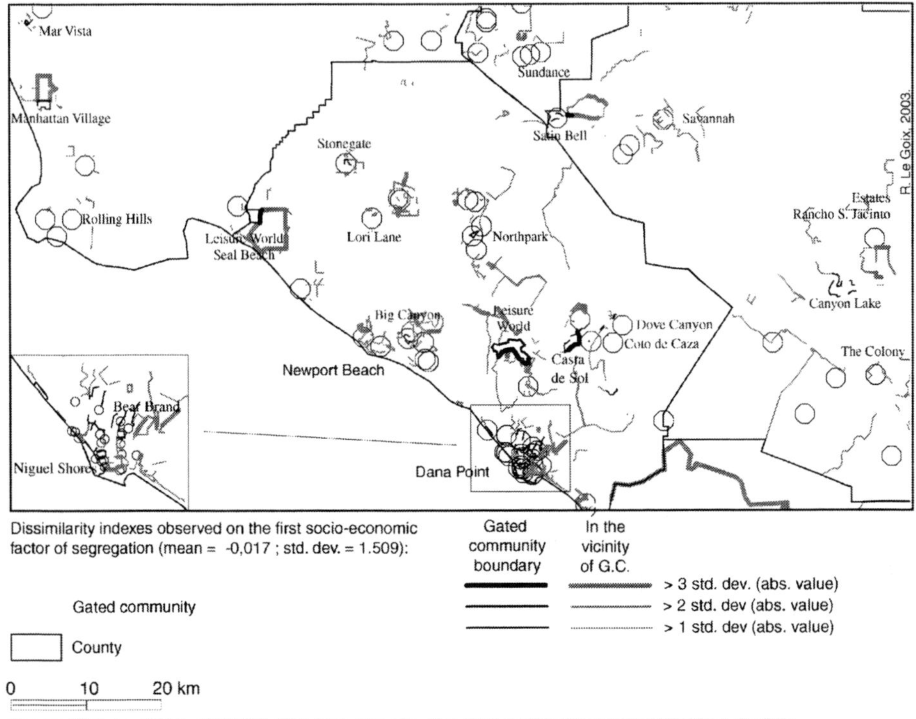

Figure 5. Major socio-economic discontinuities occurring in the vicinity of gated communities (south Los Angeles, Orange County, and west Riverside County).

a subsystem included within a larger territorial system (i.e. a municipality). Where the shapes of discontinuities are simple and circumscribe the walls, it clearly demonstrates that gated communities actually build a specific territorial system within their urban environment according to a social singularity. Within a wide range around gated communities, the shapes of the discontinuities are rather complex and depend on the shape of census block boundaries. Nevertheless, they act as evidence of major discontinuities within a certain range from the walls, thus including some gated communities within a buffer zone.

Social Walls

According to the first factor, the most relevant discontinuities can be observed around the largest retirement communities (Leisure World, Casta del Sol or Leisure Village in Ventura County), sustained by the joint effects of age, property values and white homogeneity. Partial discontinuities can also be observed along the walls (in Canyon Lake or in the Palm Springs area). As a paradox, the ones that have long been popular and documented as being prestigious enclaves do not produce strong discontinuities between themselves and their neighbours (Rolling Hills, Hidden Hills, Dove Canyon, Coto de Caza).

Although contrasts may appear at the threshold of the walls, the location of gated enclaves within a buffer zone is a common situation. In such cases, the discontinuities can be noticed within a certain distance from the wall, while the gated communities only produce weaker discontinuities. This may be observed at the north-eastern side of Dana Point, at the north side of Manhattan Village, in Newport Beach and Irvine (Big Canyon for instance), in Garden Grove, as well as in Camarillo, Calabasas and Hidden Hills (not mapped on Figure 4).

Where discontinuities around a gated community and in the surrounding areas are both significant, gated enclaves are entrenched within a double boundary. This might be explained as a buffer zone protecting the gated community from a different neighbourhood, as exemplified in Leisure World. The same method was also applied to the three other factors (although not mapped here), and it is relevant to mention that a buffer zone location usually occurs on several factors of social differentiation concurrently. Dana Point offers a good example: the major discontinuities observed on the first factor are linked to the municipal boundaries of Dana Point, whereas the major discontinuities observed on factors 2 and 4 (life cycle and age) relate to the gated community boundaries of Niguel Shores and Bear Brand Estates.

Some gated communities play a role building local social enclaves, as retirement gated communities do, but others are integrated within a larger homogeneous territory. Finally, the respective roles of gated communities and their vicinities in building local discontinuities shall be evaluated.

Segregation Factors Affected by the Enclosure

Table 1 compares the statistical distributions of dissimilarities among the different clusters in the three samples: at the level of large gated communities' boundaries (308 segments),

Table 1. Level of discontinuities observed on the first factor, in the three vicinity levels

	Frequency of discontinuities (%)				Sample parameters	
	∅	+	++	+++	Mean	SD
Factor 1: Socio-economic structure associated with race and age						
Level 1: Gated communities, boundaries	65.3	17.9	11.0	5.8	1.495	1.660
Level 2: In the vicinity	71.5	19.7	5.7	3.2	1.236	1.249
Level 3: In southern California	76.4	18.0	3.9	1.6	1.057	1.076
Factor 2: Life cycle						
Level 1: Gated communities, boundaries	66.9	14.6	9.7	8.8	1.060	1.216
Level 2: In the vicinity	74.1	18.2	5.4	2.3	0.794	0.834
Level 3: In southern California	78.7	15.3	4.0	1.9	0.710	0.782
Factor 3: Ethnic segregation effect, regardless of social status						
Level 1: Gated communities, boundaries	95.8	4.2	0.0	0.0	0.259	0.230
Level 2: In the vicinity	85.0	11.3	2.3	1.5	0.480	0.712
Level 3: In southern California	82.4	13.2	2.8	1.5	0.504	0.656
Factor 4: Age effect, regardless of social status						
Level 1: Gated communities, boundaries	60.4	17.5	4.5	17.5	1.770	2.318
Level 2: In the vicinity	76.2	15.7	4.0	4.1	0.874	1.122
Level 3: In southern California	83.7	11.6	2.6	2.2	0.662	0.890

Notes: Distributions are clustered according to mean and standard deviation of the Level 3 (33 800 segments).

Abs. values of dissimilarity indices: $\varnothing :\leq 1$ SD; $+ : 1-2$ SD; $++ : 2-3$ SD; $+++ :\geq 3$ SD

Sources: US Census 2000, block groups files SF1-SF3, database Gated Communities UMR Géographie Cités 8504, 2002.

at the level of gated communities' vicinities (6349 segments), and within the seven counties in southern California (33 800 segments).

The impact of gating is significant on factors 1, 2 and 4: major discontinuities are more frequent at the gated communities' level than in the vicinity (level 2) or in the remainder of the Los Angeles area (level 3). On factor 1, 16.8 per cent of the discontinuities rise above the two standard deviations threshold, and only 5.5 per cent in the urban region and 8.9 per cent in the vicinity of gated communities. The proportions for factors 2 and 4 are also consistent with the hypothesis of an increase of the segregation level where gates and walls are erected.

From these results, the relative increase of segregation can also be evaluated: a comparison of means of dissimilarity indices shows a higher level of discrepancy at the gates' threshold. The dissimilarities' average associated with a gated community on factor 1 is 1.4 times higher than in the remainder of the urban area, 1.5 times higher on factor 2, and 2.7 times higher on factor 4. Such a contribution of life cycle and age-based factors in the explanation of the impact of gating, though not surprising, reveals that living in a gated community is connected with age characteristics, and age homogeneity. This constitutes one of the most important factors of the social integration of those who choose to live in a gated community. This is not specific to retirement communities: everything being equal regarding the other characteristics, age seems to affect a large majority of gated enclaves as a criterion for differentiation.

Local Buffer Zones and Location Utility

Beyond the empirical evidence of a local increase of segregation spatially associated with walls and gates, it seems surprising that gated communities are not associated with race segregation. They are spatially associated with an effect two times less important on factor 3, than in the whole urban region: the average dissimilarity at level 1 is 0.259, whereas the average dissimilarity observed at level 2 is 0.480, and 0.504 at the level of the urban region (level 3). Although a paradox when considering the hypothesis usually developed about gated communities, this is demonstrated by factor 3: 95.8 per cent of gated enclaves are not associated with discontinuities based on ethnicity above the threshold of one standard deviation.

Considering the ethnic status alone, gated communities indeed always locate within very homogeneous neighbourhoods, and discriminate from their adjacent communities on the basis of age and economic status. The location within a buffer zone is not incidental, but rationally promoted by the developers choosing locations within an environment protected from the deterrent effects of ethnical diversity for the prospective buyers. This clearly affects gated communities, as far as they have to be distinctive housing for discriminating buyers concerned with the safety of their home, the security of their real-estate investment, and the social control of the urban setting. While protecting the economic value and the age-based homogeneity of the gated enclave, gated communities maximise ethnic location utility, being settled within some homogeneous ethnic environments acting as a buffer zone.

The Structure of Exclusion Around Gated Communities

A final step consists of clustering the dissimilarities observed on the four factors provided by the multivariate analysis, in order to sort out the different types of discontinuities associated with large gated communities. The underlying assumption relies on the fact that a discontinuity may occur on one factor only, or concurrently on several factors. The second possibility clearly indicates a strong structural social separation, which helps to specify the territorial identities and singularities of gated communities compared to their neighbours.

A hierarchical cluster analysis is based on the dissimilarity indices computed for 30 large gated enclaves at the level of the boundaries of the gated communities (level 1: their boundaries exactly match the census block group boundaries). The results are summarised in Table 2. The six clusters account for 78 per cent of the total variability observed.

According to these results, the following types of territorial patterns can be characterised in gated communities:

- Both clusters 1 and 2 define the retirement gated enclaves (Leisure World, Leisure Village, Casa del Sol, etc.), as the most segregated form of exclusion based on the gating. The age and life cycle are not the only predominant factors of social separation compared to the immediate vicinity, but the effects of socio-economic patterns are determinant (clusters 1 and 2), as well as the ethnic separation itself (cluster 1).
- Although the single ethnic factor is of lesser importance to explain the differentiation patterns associated with gating, four large gated communities are nevertheless associated with an ethnic segregation pattern (cluster 3). The ethnic homogeneity criterion seems to be predominant in Bradbury, in Manhattan Village, especially on its north-eastern side, where it makes contact with a more heterogeneous neighbourhood, and also in Mar Vista Gardens (Culver city), a gated public housing community where Hispanics are predominant, and which was gated according to a city security policy.
- A majority of gated enclaves produce a complex layout of discontinuities (clusters 4 and 5). They significantly differ from their neighbours by their socio-economic structure and the age factor (average profile). Nevertheless, they share boundaries with several adjacent gated communities, indeed producing a mosaic social landscape. This is true especially in Dana Point where nine major gated communities are adjacent to each other (and also in the Palm Springs area): all are rather homogeneous gated enclaves and they differentiate from each other. Complex patterns also appear in older neighbourhoods like Hidden Hills or in Rolling Hills. Hidden Hills presents an interesting case. On the eastern edge, it shares a boundary with the city of Los Angeles and the discontinuity is close to the average profile (cluster 5); on its southern boundary with the city of Calabasas the discontinuity is based on ethnical differentiation (cluster 3); on its western edge no noticeable social discontinuity can be observed with the recent upper-scale gated enclave of Mountain Gate.
- Some gated communities do not differentiate more than the average profile, and are almost integrated with their environment (cluster 5), like Canyon Lake, Dove Canyon and Coto de Caza (Figure 5). An interesting case indeed: in 2000, while Dove Canyon was incorporating with the rest of the city of Rancho Santa Margarita, Coto residents opted out of the incorporation on the argument they have different socio-economic profiles (Ragland, 1999; Tessler and Reyes, 1999). It is always interesting to compare what the residents of gated communities think

Table 2. Cluster analysis of the dissimilarities observed between the gated communities and their immediate vicinities (hierarchical clustering, 308 spatial units, 6 clusters and 78% of variability explained)

Cluster (and tree)	Freq.	Contribution of the variables (dissimilarities observed for each factor)			
		Factor 1: Socio-economic structure associated with race and age	Factor 2: Life cycle	Factor 3: Ethnic segregation, regardless of social status	Factor 4: Age effect, regardless of social status
1. Complete territorial discontinuity on all the factors	1.6%	+++	+++	++	+++
2. Retirement enclaves with ethnic homogeneity	15.3%	++	++		++
3. Enclaves based on ethnic discontinuity	14.6%			++	
4. Enclaves similar to the average profile	33.1%		(+)	(−)	
5. Enclaves below the average profile	24.0%	(−)	(−)		(−)
6. Poorly defined discontinuities	11.4%	–	–	–	–
Average profile: compared to the discontinuities observed in the whole Los Angeles area, the gated communities are locally associated with an average discontinuity...		1.4 times higher	1.5 times higher	2 times less	2.7 times higher

Notes: +/−: contribution ≥ +/−0.5 SD; ++: contribution ≥1 SD; +++: contribution ≥2 SD
(+) and (−): low contribution <0.5 SD
Sources: US Census 2000, database Gated Communities UMR Geographie-cités.

of themselves, as it is highly connected with a 'snob' value and a subjective 'distinction', although there is no socio-economic distinctive pattern outlined by a consistent discontinuity (Figure 5).

- A final group includes the large gated communities where no major discontinuity (cluster 6) can be observed (Canyon View Estates in Santa Clarita and Big Canyon in Irvine). In these two cases, the gate does not separate different social groups and only emphasises the private property and the exclusiveness of the amenities.

The typology highlights the variety of the insertion of gated communities within their neighbourhoods. The five categories of gated communities demonstrate that gates and walls contribute to the spatial integration of social territories. The statistical significance of the phenomena (although some bias was discussed) acts as evidence of the construction of gated communities as homogeneous and differentiated territorial systems that intensify segregation at a local scope. These results are of interest because they better qualify the nuances occurring in segregation patterns where gated communities are present. First, the effect of age always determines the singularity of gated communities compared to their local surroundings, even for gated communities that are not retirement communities. This might suggest that middle-aged people and seniors are both attracted by the developers' discourses about security their willing as homeowners to protect a lifetime investment, and gated communities are an efficient answer to those concerns. Second, the combined effects of property values and socio-economic structure of the population (factor 1) suggest that the homeowners usually consent to a higher level of investment in gated neighbourhoods than in open neighbourhoods in the surrounding areas. Finally, the effect of ethnicity must be analysed carefully: it does not contribute at all to define gated communities as 'worlds apart', except in a few cases like Bradbury and Manhattan Village. Nevertheless, gated communities always locate within homogeneous areas (on ethnic criteria) that act as buffer zones protecting the enclave from heterogeneous neighbourhoods by a thick 'wall of ethnic homogeneity'. In fact, the municipalities where gated communities are settled often fit the limits of this 'buffer zone' of homogeneous people. Gated communities stress an exclusion that is also structured by public policies at a municipal scale.

Conclusion

The analysis of gated communities as territorial systems defined by a physical and juridical border is of interest to understand their local impact and the spillover effects they might produce because of their numerous interactions with local governments and the social environment.

First, this focus highlights the originality of gated communities in the suburban development that depends on the enclosure rather than on an architectural singularity of the neighbourhood itself. A gated community is nothing but a Common Interest Development, and often looks like some other neighbourhoods in the surroundings. Nevertheless, the enclosure favours the property value and increases the property tax basis. Furthermore, the erection of gates transfers the cost of maintaining the urban infrastructure to the association and the homeowner. The relationships between the gated enclaves and the public authorities can be thus summarised: because of the fiscal basis they produce, at almost no cost except general infrastructures (freeways and other major infrastructures), gated communities are particularly desirable for local governments. The sprawl of gated communities is not to be understood as a trend towards a 'secession' from the public authority but as a public-private partnership, a local game where the gated community has a financial utility for the public authority, whilst the property owners' association is granted more autonomy in local governance, as discussed in the case of Calabasas (for critical material regarding this issue, see also Le Goix, 2003). This ambiguous relationship helps to get a better understanding of the reasons leading to a sprawl of gated communities that cannot be simply explained by a rush for security.

Second, the analysis seeks to demonstrate how the gated territorial construction produces spillover effects. Not only do gated communities probably divert crime and protect property values (with a deterrent effect for property values in the surroundings), but it was possible to measure the socio-economic effects of this structuring of urban space at a local scale: gated communities are homogeneous territories that differentiate from their neighbours especially on age criteria and socio-economic status. A final conclusion highlights the strategies of developers: gated communities are often located within a buffer zone of homogeneous ethnic patterns, and these buffer zones often fit the municipal boundaries. Gated communities do not increase segregation on their own. They belong to a process of production of urban space made by private strategies (the developers) and public strategies (attracting taxpayers) which is finally consistent with the long involvement of public policies with segregation processes, as Massey and Denton (1993) pointed out in the US. The diffusion of gated communities is not only supported by developers and homebuilding industries, but also by public authorities earning their share in the process.

Acknowledgements

This paper is drawn from a doctoral thesis founded by the CNRS (UMR Géographie-cités 8504, Paris), the French-American Foundation (Tocqueville Fellowship, 2000–01), and the French-American Commission (Fulbright Research Scholarship, 2002–03), which is gratefully acknowledged. The author also sincerely wishes to thank Pr. Allen J. Scott (UCLA) and Pr. Setha Low (CUNY) for their advice, comments and corrections to an earlier version of this paper.

Part of an earlier version of this paper was presented at the International Conference on Gated Communities, Glasgow, 18–19 September 2003.

References

Apparicio, P. (2000) Indices of residential segregation: a tool integrated in geographical information systems, *Cybergeo*, 134. Available at http://cybergeo.presse.fr.

Bator, F. M. (1958) The anatomy of market failure, *Quarterly Journal of Economics*, 72, pp. 351–379.

Bible, D. S. & Hsieh, C. (2001) Gated communities and residential property values, *Appraisal Journal*, 69, p. 140.

Blakely, E. J. & Snyder, M. G. (1997) *Fortress America: Gated Communities in the United States* (Washington DC and Cambridge, MA: Brookings Institution Press & Lincoln Institute of Land Policy).

Brown, K. (1991) *Mello-Roos Financing in California* (Sacramento, CA: State of California, California Debt Advisory Commission).

Brunet, R. (1965) *Les phénomènes de discontinuité en géographie* (2nd edn, 1970) (Paris: Editions du CNRS).

Caldeira, T. P. R. (2000) *City of Walls: Crime, Segregation, and Citizenship in Sao Paulo* (Berkeley, CA: University of California Press).

Castells, M. (1983) *The City and the Grassroots* (Berkeley, CA: University of California Press).

Ciotti, P. (1992) Forbidden city, *Los Angeles Times*, 9 February, p. B3.

Curtin, D. J. (2000) *California Land-use and Planning Law, 2000* (Point Arena, CA: Solano Press Book).

Davis, M. (1990) *City of Quartz: Excavating the Future of Los Angeles* (London: Verso, coll. The Haymarket Series).
Davis, M. (1998) *Ecology of Fear: Los Angeles and the Imagination of Disaster* (New York: H. Holt).
Dear, M. & Flusty, S. (1998) Postmodern urbanism, *Annals of the Association of American Geographers*, 88, pp. 50–72.
Decroly, J. M. & Grasland, C. (1992) Frontières, systèmes politiques et fécondité en Europe, *Espace, Populations et Sociétés*, 2, pp. 135–152.
Donzelot, J. (1999) La nouvelle question urbaine, *Esprit*, 258, pp. 87–114.
Donzelot, J. & Mongin, O. (1999) De la question sociale à la question urbaine, *Esprit*, 258, pp. 83–86.
Flusty, S. (1994) *Building Paranoia: The Proliferation of Interdictory Space and the Erosion of Spatial Justice* (West Hollywood, CA: Los Angeles Forum for Architecture and Urban Design).
Foldvary, F. (1994) *Public Goods and Private Communities: the Market Provision of Social Services* (Aldershot: Edward Elgar).
Forsyth, A. (2002) Who built Irvine? Private planning and the federal government, *Urban Studies*, 39, pp. 2507–2530.
Fox-Gotham, K. (2000) Urban space, restrictive covenants and the origins of racial segregation in a US city, 1900–50, *International Journal of Urban and Regional Research*, 24, pp. 616–633.
Francois, J.-C. (1995) Discontinuités dans la ville. L'Espace des collèges de l'agglomération parisienne (1982–1992), Doctorate thesis, geography, Université Paris 1-Panthéon-Sorbonne.
Frey, W. H. (1993) The new urban revival in the United States, *Urban Studies*, 30, pp. 741–774.
Garreau, J. (1991) *Edge City: Life on the New Frontier* (New York: Doubleday).
Glasze, G., Frantz, K. & Webster, C. J. (2002) The global spread of gated communities, *Environment and Planning B: Planning and Design*, 29, pp. 315–320.
Grasland, C. (1997) L'analyse des discontinuités territoriales. L'exemple de la structure par âge des régions européennes vers 1980, *L'Espace Géographique*, 26, pp. 309–326.
Helsley, R. W. & Strange, W. C. (1999) Gated communities and the economic geography of crime, *Journal of Urban Economics*, 46, pp. 80–105.
Jackson, K. T. (1985) *Crabgrass Frontier: The Suburbanization of the United States* (Oxford: Oxford University Press).
Jaillet, M.-C. (1999) Peut-on parler de sécession urbaine à propos des villes européennes?, *Esprit*, 11, pp. 145–167.
Kazmin, A. L. (1991a) 5 Council Members are sworn in as Calabasas marks debut as a city, *Los Angeles Times*, 6 April, Metro, part B, p. 4.
Kazmin, A. L. (1991b) Calabasas cityhood appears a certainty, *Los Angeles Times*, 24 February, Metro, part B, p. 3.
Kennedy, D. J. (1995) Residential associations as state actors: regulating the impact of gated communities on non-members, *Yale Law Journal*, 105, décembre.
Knox, N. H. & Knox, C. E. (Eds) (1997) *California General Plan Glossary* (Palo Alto: CA California Planning Roundtable, the Governor's Office of Planning and Research).
Lacour-Little, M. & Malpezzi, S. (2001) *Gated Communities and Property Values* (Madison, WI: Wells Fargo Home Mortgage and Department of Real Estate and Urban Land Economics, University of Wisconsin).
Le Goix, R. (2001) Les 'communautés fermées' dans les villes des Etats-Unis: les aspects géographiques d'une sécession urbaine, *L'Espace Géographique*, 30, pp. 81–93.
Le Goix, R. (2002) Les gated communities en Californie du Sud, un produit immobilier pas tout à fait comme les autres, *L'Espace Géographique*, 31, pp. 328–344.
Le Goix, R. (2003) Les 'gated communities' aux Etats-Unis, morceaux de villes ou territories à part entière? (Gated communities within the cities in the US: Urban neighborhoods or territories apart?), Doctorate thesis, geography, Université Paris Panthéon-Sorbonne. Available at http://tel.ccsd.cnrs.fr/documents/archives0/00/00/41/41/index_fr.html.
Low, S. M. (2001) The edge and the center: gated communities and the discourse of urban fear, *American Anthropologist*, 103, p. 45.
Marcuse, P. (1997) The ghetto of exclusion and the fortified enclave: new patterns in the United States, *The American Behavioral Scientist*, 41, pp. 311–326.
Massey, D. S. & Denton, N. A. (1993) *American Apartheid: Segregation and the Making of the Underclass* (Cambridge, MA: Harvard University Press).
McKenzie, E. (1994) *Privatopia: Homeowner Associations and the Rise of Residential Private Government* (New Haven and London: Yale University Press).

Miller, G. J. (1981) *Cities by Contract* (Cambridge, MA: The MIT Press).

Moore, M. (1995a) Brentwood Road block communities: residents of enclave near the Getty Center receive tentative City Council approval to erect gates, *Los Angeles Times*, 18 May, p. 3.

Moore, M. (1995b) Part of Brentwood allowed to become gated community, *Los Angeles Times*, Home Edition, 4 June, p. 3.

N'Guyen, T. (1999) Coto de Caza residents say no to school within gates, *Los Angeles Times*, Orange County Edition, 4 March, Sect. B, p. 1.

Pool, B. (1987a) Calabasas cityhood backers dealt another major setback, *Los Angeles Times*, 1, 1 May, Metro, part 2, p. 6.

Pool, B. (1987b) Calabasas cityhood bid falters again; 3 residential areas, industrial park omitted; new finance study ordered, *Los Angeles Times*, Valley Edition, 6 April, Metro, part 2, p. 10.

Pool, B. (1987c) Mountains or molehills? Calabasas cityhood backers contest builder's right to exclude 1,300 acres of ranchland, *Los Angeles Times*, 29 August, Metro, part 2, p. 6.

Purcell, M. (1997) Ruling Los Angeles: neighbourhood movements, urban regimes, and the production of space in southern California, *Urban Geography*, 18, pp. 684–704.

Ragland, J. (1999) City-to-be, county agree on tax deal, *Los Angeles Times*, 23, June, p. 5.

Reich, R. B. (1991) Secession of the successful, *New York Times Magazine*, p. 16.

Sanchez, T., Lang, R. E. & Dhavale, D. (2003) *Security Versus Status? A First Look at the Census's Gated Communities Data* (Alexandria, VA: Metropolitan Institute, Virginia Tech).

Scott, A. J. (1980) *The Urban Land Nexus and the State* (London: Pion).

Sorkin, M. (1992) *Variations on a Theme Park: The New American City and the End of Public Space* (New York: Hill and Wang).

Stark, A. (1998) America, the gated? (impact of gated communities in political life), *Wilson Quarterly*, 22, pp. 50–58.

Tessler, R. & Reyes, D. (1999) 2 O.C. Gated Communities are latest to seek cityhood, *Los Angeles Times*, 25, January, p. 1.

Webster, C. J. (2002) Property rights and the public realm: gates, green belts, and Gemeinschaft, *Environment and Planning B: Planning and Design*, 29, pp. 397–412.

Gated Communities as Club Goods: Segregation or Social Cohesion?

TONY MANZI & BILL SMITH-BOWERS
London Research Focus Group, University of Westminster, School of Architecture and the Built Environment, London, UK

(Received October 2003; revised May 2004)

KEY WORDS: Gated communities, residential segregation, club goods

Introduction

The issue of gated communities raises important questions about the future forms of urban development. In much of the academic literature the proliferation of gating is treated as an indicator of increasing levels of social division; creating new barriers between rich and poor, and introducing 'cities of walls' (Brunn *et al.*, 2003; Caldeira, 2000; Sandercock, 2002; Scott, 2002). The standard perception of gated communities is that design and technological innovations serve to increase privatism and destroy traditional community ties of neighbourliness, community and cohesion (Gottdiener & Hutchison, 2000).

The notion that gating exclusively benefits an elitist minority forms a deep-rooted belief in much of the literature. For example, Joseph Rykwert (2002) describes some of the recent additions to the Manhattan skyline (such as the Trump World Tower) as "vertical gated communities" offering "a commanding residence for the privileged few" (p. 218).

What is a gated community? An influential publication offers the following definition:

> Residential areas with restricted access in which normally public spaces are privatised. They are security developments with designated perimeters, usually walls or fences, and controlled entrances that are intended to prevent penetration by non-residents. (Blakely & Snyder, 1997, p. 2)

Gating therefore involve an inevitable form of privacy and exclusivity. Moreover, the stereotypical view of gated communities is that they embody a form of dystopian living, behind which community ties are non-existent with neighbours discouraged from developing social interactions. In particular, they are seen to encourage affluent groups to increase their social distance from what is perceived as the 'other'. A common representation of gating is derived from Davis' (1990) *City of Quartz*, where the concept of 'Fortress America' encapsulates an increasing polarisation between rich and poor in cities such as Los Angeles. Davis contends that "we live in 'fortress cities' brutally divided between 'fortified cells' of affluent society and 'places of terror' where the police battle the criminalised poor" (p. 224). Davis' thesis is deliberately polemical, but nevertheless highly influential in constructing a negative image of the gated society. Hence:

> A pliant city government ... has collaborated in the massive privatisation of public space and the subsidisation of new, racist enclaves (benignly described as 'urban villages') ... a triumphalist gloss ... is laid over the brutalisation of inner-city neighbourhoods. (p. 227)

Although rarely described in such stark dichotomies—Davis refers, for example, to 'spatial apartheid' and a 'Berlin wall' separating 'publicly subsidised luxury' from a 'lifeworld' 'reclaimed by immigrants' (p. 230)—these fears have permeated the policies of inner-city local planning authorities. Central and local governments in the UK have therefore attempted to prevent a replication of the spatial polarisation of North American inner cities, by discouraging gated developments, restricting planning approval and encouraging neighbourhood renewal schemes based on more 'traditional' design layouts (ODPM, 2000).

Others argue that gating is a feature of the growth of 'global city regions' and the intensification of inequality and proximity which has accompanied urban growth and globalisation of the 'free market':

> Violence, or the fear of it, becomes a central preoccupation of the upper classes, pushing them towards forms of fortress settlement, gated high-rise communities surrounded by walls and guarded entries. (Scott, 2002, p. 25)

Gated communities are thus seen as a feature of growing importance in the development of residential segregation taking place within cities. Some writers suggest that gating is an overreaction to the real level of crime in an area compared to the perceived level of crime that results from local media coverage of crime incidents in the USA. This argument is part of the 'culture of fear' thesis put forward by Glassner (1999) suggesting that fear of crime is just one of a number of 'panics' (that also include deadly diseases, teenage lone mothers and African American males) propagated by local television news and current affairs programmes. An over-emphasis on individual cases results in unnecessary risk reduction responses to these events. Glassner argues that the underlying drives of many of the current problems of American cities, such as poverty and income inequality are neglected: "One of the paradoxes of a culture of fear is that serious problems remain widely ignored even though they give rise to precisely the dangers that the populace most abhors" (p. xviii).

The 'culture of fear' is explained as the result of people embracing 'improbable pronouncements' (his example being the response of many Americans to the broadcast of Orson Welles' *War of the Worlds* in 1938). Glassner suggests that acceptance of these 'pronouncements' is the result of how they are delivered by 'professional narrators' and presented in news and current affairs programmes as 'statements of alarm', "poignant anecdotes in place of scientific evidence, the christening of isolated incidents as trends, depictions of entire categories of people as innately dangerous" (p. 208).

Many approaches to the phenomenon of gating suggest that it is a response to increasing social inequalities, status-seeking behaviour, and real or perceived fear of crime (Lindstrom & Bartling, 2003). References to the 'totalitarian semiotics' (Davis, 1990, p. 231) of urban design mark a deliberate attempt to deny the validity of certain forms of development per se. Consequently, rather than allowing local preferences to shape decision making (as is claimed by many such critics), such analyses presume that gating by definition is a form of design that should be rejected out of hand. Thus, heterogeneity is acceptable as long as it does not result in a denial of public space. Is this commitment to the public realm to be defended at all costs?

Club Goods and Gated Communities

An alternative approach to sociological and anthropological analyses of gated communities can be found in the economic literature on clubs. The notion of a 'club good' originates in the work of Buchanan (1965), who uses it to examine jointly consumed and excludable services. Buchanan argued that there is a type of good (the club good), which like private goods had excludable benefits but was allocated through groups. This allowed the club members the enjoyment of the benefit but was unlike the private good which is limited to the individual or shared by all in the case of the public good. The club good is neither a 'private' nor 'public' good in the traditional economic sense. Rather, it constitutes a hybrid in which a self-selecting community shares a range of benefits and reduces the costs of public good 'congestion' by the use of its pricing and membership requirements. Buchanan introduced the concept of a continuum of ownership and consumption possibilities which created a bridge between the purely public and the purely private goods that had formerly been the focus of discussion in the economic literature (see Samuelson, 1954; Tiebout, 1956).

Gated communities can be analysed in economic terms as a form of holding property rights developed through collective action of individuals for individual and mutual benefits. This makes gated communities different from private goods, such as a single dwelling owned by a person, and public goods, such as a public park, because while there is sharing of benefits (which is the definition of a public good) there is also 'excludability' of benefits (the definition of a private good). The hybrid quality of this good sharing has led to the concept of club economics being used to explain this type of commodity.

Chris Webster has extended the concept of club goods to the analysis of gated communities (2001, 2002). Webster was the first to point out that the Garden City plans of Ebenezer Howard were in reality plans to develop a private city as a club good (2001). The reconceptualisation of this concept of collective action to secure club goods was further developed and applied to gated communities across a range of different societies (Webster & Wu, 2001; Webster & Wai-Chung Lai, 2003). This work focuses on the management of property rights and uses the concept of 'proprietary communities' to delineate the nature of the gated community. The gated community development thus provides desired goods and services such as "security zones, lifestyle, retirement and prestigious communities" (Blakely & Snyder, 1997, pp. 38–45). In club economic terms gated communities are merely a recent example of the growth of privately owned collective goods such as shopping malls, business parks, timeshare apartments, golf and squash clubs.

Developing Webster's argument, it is suggested that households are willing to purchase different forms of rights in securing their accommodation and communal service requirements. At the start of the 20th century, most households exercised rights associated with renting or long leasing a part of a property. During the second half of the 20th century, the trend was for more and more households to purchase the rights associated with the ownership of freeholds and entire properties. By the 21st century, we are witnessing the growth of gated communities because the additional rights and obligations of this desired and scarce good are now being priced competitively for more households. When purchased, these can enhance the traditional benefits associated with freehold or leasehold occupation. Therefore, gated communities offer a range of scarce goods, such as secure and guaranteed parking, enhanced security, common standards for property appearance and rules governing the use of managed communal areas.

Furthermore, whilst formerly associated with elite groups who could afford the luxury of these kinds of purchases, rising real incomes and the comparative fall in security and monitoring costs are bringing these goods within the budgets of middle-income households. In contrast to much academic commentary, recent research from the USA by Sanchez & Lang (2002) suggests that the view of gated communities as the preserve of the white high-income homeowner is exaggerated. Their analysis of the 2000 census (which included for the first time questions on gated communities) identified significant numbers of poorer white and ethnic minority renters who live in gated communities. They conclude that gating not only functions as a status symbol for the better off homeowners, but also provides a response to fear of crime and protection for lower-income renters.

In addition to its physical and environmental attributes, private communal areas, walls, gates and security patrols, the gated community constitutes a 'territorial organisation' of the community members' property rights (Glasze, 2003; McKenzie, 1994). These can include homeowners associations (HOAs) or Common Interest Housing Developments (CIDs) (McKenzie, 2003). In principle, these organisations provide a vehicle of representative government in the management of community interests. Both Glasze and McKenzie have questioned how democratic and representative such associations are in practice. However, the additional merit good of being able to directly influence the management of a community is one of the key objectives of the UK government's neighbourhood regeneration policy (DETR, 1998). Furthermore, the concept of choice has become an increasingly important aspect of housing service delivery (ODPM, 2000).

Whilst there are significant differences between the US and UK environments, a central preoccupation of policy debate has been a focus upon housing design, which has been long recognised as having a huge impact on the way in which public spaces are used and perceived. Newman (1972) and Coleman (1985) advocated 'territorialism' and making the public realm 'more defensible' in the interest of safety, long before gated communities became an issue of public policy in the UK. A primary motivating factor in the growth of the 'gated' community and the 'alley-gating' phenomenon in the UK, as in most other countries, has been the rise in both public concern and government concern about crime, vandalism and anti-social behaviour (Garland, 2001). Some people, for example vagrants, drug users and gangs of young people are perceived as causing conflict within the public realm and often attempts are made to 'design them out' through the use of gates, CCTV cameras and other physical barriers (Raco, 2003). Often, however, this can also have the effect of closing off the space to the general public.

Other concerns linked to access to parking spaces in London and protection of vehicles have promoted a significant growth in gated developments. These concerns have also occurred at an important time in the development and falling costs of some types of security devices and their incorporation in the design of buildings. The gated option for individuals, property developers and social landlords is now cheaper and more feasible than ever before.

This issue is raised by the current alley gating programmes which have been developed using regeneration funding in areas as far apart as Manchester, Liverpool and Watford. Because of the evident support it has generated among local residents in poorer neighbourhoods (see Landman, 2003; Mumford & Power, 2002) the policy is currently to encourage alley gating. Thus, in July 2003, the government announced that local authorities would have the power to close rights of way in certain blighted areas in order to reduce the opportunity for criminal activity (DEFRA, 2003). While this may reduce burglar access to properties inside these gates, in many cases it will also prevent the continued use of these alleyways as safe pedestrian routes to local services.

Thus, while government policy rejects the gating of streets and the creation of gated communities (ODPM, 2000) it supports the gating of alleys (small streets) and the creation of gated areas, in many cases removing traditional rights of way, which planners argue should be one of the reasons to reject the creation of gated communities. This marks a contrast with other areas of government policy to open up rights, for example, in the Countryside and Rights of Way Act (2000).

This paradox is also reflected in other discussions on gated communities (for example RICS, 2002). The RICS report both expressed concern at the problem of social segregation and lack of planned growth of gated communities and concluded: "Policies to create greater balance should be directed towards new development, which increasingly includes gated communities, as well as the regeneration of blighted areas" (p. 6).

If security, exclusive use of communal services, the managed prevention of unsolicited calling and guaranteed parking are valued by community members, the key issue raised by gated communities is who can enjoy these benefits and are some households socially excluded from these benefits? This is not a new argument; it arose at the beginning of the 20th century when governments commenced providing rented housing as a merit good at below market price to selected households. The debate evolved in the 1980s to encompass the additional promotion of owner occupation via the Right to Buy provisions of the 1980 Housing Act and the emergence of shared ownership and other mechanisms for promoting ownership among lower-income households.

The question today is whether or not gated communities can be regarded as a merit good with public subsidy offered to enhance the provision and enjoyment of that good and service. This argument is not hypothetical because in the UK the state subsidises gated and managed sheltered accommodation for older people as well as alley gating in areas of high crime.

Residential Segregation and Gated Communities in a UK Context

Social relations and social interactions within public housing space are fundamentally determined by the people who live there alongside a wider process of market and social housing allocation. In this respect, the locality and nature of housing is a major determinant of how connections between individuals and communities are formed and maintained. It is generally accepted that the distribution of residential units and their occupants is not a consequence of random events but the product of complex social, economic and political processes. One of the most significant results of these processes is that housing consumption patterns can result in segregated areas otherwise known as 'enclaves' (suggesting choice) or 'ghettos' (suggesting constraint).

Part of the difficulty is that segregation is a highly loaded term. As Smith (1989) acknowledges, "Segregation in its broadest sense refers to the organisation of all social life. It has to do with the conditions of interaction or avoidance, the construction of group identity and the structuring of social, economic and political life" (p. 14). Any process that increases residential segregation is therefore viewed with outright hostility by most commentators, following the Chicago School of spatial sociology, whereby positive behaviour and attitudes are generated by removing distance and increasing interaction (Miller & Brewer, 1984; Smith, 1989, p. 14).

The economic theory of 'collective action' and 'club goods' has also been applied to the concept of social exclusion in the work of Jordan (1996). The group and its selective membership criteria promote acting in a way not detrimental to the group interest, thus maximising the benefits to members while reducing or preventing benefits accruing to non-group members. Jordan provides the example of the UK National Health Service, which cannot provide customer services that higher-income groups seek and who therefore join health insurance schemes that screen out those with poor risk from membership, thus intensifying social exclusion in terms of access to health care linked to income and health inequalities. If applied to gated communities, Jordan's argument would view these as another type of club that creates a new form of spatial social exclusion.

However, as housing has never been viewed as a good to be supplied free at the point of access (in the same way as the NHS), it can be argued that the gated community functions as a merit good in which choice can play a crucial role; thus public subsidy can be applied to increase safety and security in regeneration areas.

There is a lack of empirical research examining the consequences of gated developments within a UK context. Far greater evidence exists on the impact of 'gated' communities within the US literature (for example, Blakely & Snyder, 1997; Low, 2003). A systematic review of literature in the UK found little discussion of the implications of having developments where residents segregate themselves from the perceived threats of the outside world (Blandy *et al.*, 2003). Other studies have been commissioned which have argued that the growth of gating is a serious threat to community cohesion and urban sustainability through spatial and temporal segregation, in addition to conflicting with English urban cultural traditions (Atkinson *et al.*, 2003, p. 5). Whilst there is little specific guidance on gated developments, government policy places strong emphasis on social cohesion achieved through interaction and contact between different social and cultural groups (Home Office, 2001; ODPM, 2000). The general assumption appears to be that gated developments detract from these objectives.

Methodology

The research conducted for this paper consisted of case studies of two gated developments. The aim of the research was to understand how gated developments were perceived by residents in different environments and to provide qualitative empirical evidence about the impact of gated communities upon distinctive urban environments. This study originated in a wider research project on an outer London estate into neighbourhood profiling and neighbourhood renewal. The client of the project was not primarily interested in the gated community (see Manzi & Smith-Bowers, 2004). However, the authors considered this gated community raised important issues about reducing spatial segregation and the types of developments that should be provided on mixed tenure estates. A second 'gated community' was selected in inner London, which was in the process of development; this was also located in an area of multiple deprivation as defined by the ODPM index.

The research in outer London used interviews with residents of the estate, the chair of the residents' association of the gated community, local landlords and service providers, observation of forum events and community meetings. In the inner London development interviews were conducted with the managing agent of the estate, the security consultant, the planning officer and residents. The Outer London scheme was located within a mixed tenure estate and the other was designed as a private development (with additional social housing to be provided at a later stage). The former was a permanent gated settlement and the latter a temporary gated environment. The initial purpose of the interviews was to gather more detailed information about management issues, relationships in the neighbourhoods, local service delivery and priorities for improvement. A total of 20 interviews were conducted.

Case Study One: The Permanent Gated Community

The first development is a mixed tenure estate in outer London built in the first half of the 1990s. The estate, which comprises one-third of the ward population, is located in a neighbourhood ranked 634 out of 8414 on the Index of Deprivation (DETR, 2000). The estate is divided into a number of sub-sections. A wall with two electric gates to permit and restrict entry to residents and their guests separates the owners from the wider estate. This part of the development houses around 200 owner occupiers in the converted wing of a 19th century asylum. The remainder of the estate is situated outside the gated area and is semi-enclosed within the historic walled grounds of the 19th century asylum (but without gates). In this part of the estate about 600 units of social housing, shared ownership and private renting accommodation are located in different sub-developments. The estate can be described as a 'forted up' mixed tenure development inside two sets of walls.

These walls and gates were considered a key problem within this development in that the social housing estate is physically separated from the privately owned and gated community. One local authority officer expressed the difficulty in the following terms:

> It has a history as a psychiatric hospital ... I see it as the final bastion of stigmatisation. It reinforces the sense that it is still a madhouse; it is symbolic of care in the community. You put them in houses and put a wall around them. It conspires with a subliminal message ... You could believe that it is still a psychiatric hospital. You should not underestimate the symbolism of the physical. Walled cities in ancient times were fortresses to keep people in and out. The physical fabric is testimony to separateness. (Interview)

The estate was the largest RSL consortium development in the UK of the early 1990s and probably the only one to contain a gated community within its boundary. From its start the estate brought together many contemporary features of housing development, private ownership and leasing, shared ownership and social renting, RSL consortium development and a gated community (only local authority housing is absent from the landlord mix). In one sense the estate was a leading example of a mixed community development, in that it brought a range of income groups together in one neighbourhood rather than being segregated into different residential neighbourhoods.

However, the practicalities of mixing diverse social groups proved highly problematic. The development was not planned as a social housing scheme and much of the infrastructure planned did not materialise (Interview data). In addition, from the beginning there was a strong feeling of segregation between social housing residents on the one side and private owners and leaseholders on the other. As one private resident commented: "there was a real 'us and them' scenario" (Interview). This meant that owners and leaseholders did not see themselves as benefiting from the community facilities:

> I very rarely go to the ... shop. They can tell you by the car you drive or the way that you dress ... that you are not from the housing association flats. It is aggressive. (Interview)

A strong sense of conflict was generated between the different social groups on the estates. This was expressed in the following way by a leaseholder in one of the flats which was located on the estate but not within the gated community: "there is definitely a bad feeling towards the people living in these flats because we are owners. There is a definite class divide I think" (Interview).

The owner occupiers within the gated community also felt removed from much of the day-to-day activities on the estate. Because they did not share the experience of the majority of residents in the neighbourhood the scale of the social problems reported by other residents surprised them. For example, one owner occupier commented:

> I have been to a few of the resident meetings. We were absolutely horrified to hear what they were saying about prostitution and drug abuse. Residents said that they knew who was perpetrating these crimes but that they did not dare come forward to report them due to the fear of reprisals. I also heard that some of the neighbours did not come to the meeting as they were watching who was attending. It was felt that it was a 'grassing' situation. (Interview)

In addition, the gated residents were aware of the class distinctions between those within and outside of the walled community and acknowledged that a high level of diplomacy was called for in making contributions to collective management.

> I am the only one who has gone to the ... meetings. I am very careful about what I say. I know that a lot of them are on income support. For example if I talk about kids damaging our cars, I need to be diplomatic. You only have to compare the cars inside and outside. (Interview)

Despite the disparities in income and wealth, there appeared to be some co-operation between residents; in particular they felt they shared common goals in terms of improving their neighbourhood. Nevertheless, residents felt that the gated development was essential in preserving a sense of security and protection from the varied social problems occurring on the estate.

> A couple of people were mugged ... when they were waiting for the gates to open. It was a prime opportunity as they had to get their swipe cards from their wallets. We used to have a code to enter the grounds but [the youths] knew the code. They are not stupid. I dread to think how much we are paying for the gates but they are a necessity. When they were broken (by the kids of course) cars were getting broken into. (Interview)

Despite the very serious social problems on the estate, voiced by residents and workers in the neighbourhood, owners generally felt happy and secure in their properties.

> I bought the flat at a very good price. I have never felt unsafe inside. I have installed a spy hole and extra window locks. For the first two years I lived on my own. The gates have done a lot to help. Personally I have never had problems that I wouldn't find on any London street but I tend not to walk around the estate. (Interview)

Such views illustrate how there can be reasonable levels of safety and security despite residents living within an area widely perceived as a high crime neighbourhood. Significantly, there appeared to be very different perceptions between those within the gated community (who were largely positive) and those living in leasehold flats that were integrated within the social housing estate. The latter appeared much more negative about their environment and reported much more serious instances of harassment, intimidation, victimisation and crime.

As argued above, the gated community is not normally identified as one of the aspects of a mixed community development in the statements of government and other interested parties. Rather it is commonly viewed as the opposite of a desirable social mix in urban living the government wishes to promote; gated communities challenge these aspirations given their target population of affluent households. However, the legal structure means that most are owned and managed collectively by the residents. This represents a further issue of collectivism versus individualism, given that one of the 'solutions' to the sustainable development of the estate was seen as the development of tenant management. Such trends represent what can be termed "an unusual blend of collectivism combined with a retreat into privatised spaces" (Blandy *et al.*, 2003, p. 3). This demonstrates how the phenomenon of gating can be connected to club goods theories to illustrate new approaches to private and public service provision. As Webster (2001) maintains it is misleading to polarise debate into issues of public versus private institutions.

This case study suggests that one way to promote mixed tenure developments in areas of deprivation is to acknowledge community members' concerns for safety and security. The study suggests this can be done by developing gated sub-subsections in the neighbourhood.

Case Study Two: The 'Lifestyle' Temporary Gated Community

Owned by a large private sector property development company, this southeast London site was previously a derelict industrial estate. The development is located in one of the poorest wards in the country, ranked 468 out of 8414 on the Index of Deprivation (see DETR, 2000). The development is an example of the vision of the local authority to use culture and the arts as a driver to regenerate the area and bring higher-income households into the inner city. It also meets the objectives of the economic regeneration strategy of the borough to create accommodation for office workers. The estate manager explained the developer's objectives:

> The vision was to design a 'new concept for living'—a 'lifestyle' community. This encapsulates a total living environment comprising home and leisure facilities with 24-hour concierge service to care for residents' every requirement. (Interview)

The advertisements and marketing for the scheme present the development as a prestigious housing and living complex situated in what could be taken as an upmarket area. However, the immediate location is not the focus of the marketing of the estate. The main selling points of the area are the local rail station opposite the development and the lifestyle that was available inside the complex at affordable prices. The marketing focus was on the 'living experience', referring to modernist interiors and immediate surrounding exterior facilities such as a gym, landscaping and restaurant. It was presented in publicity material as "the development where you *can* have it all".

The development was targeted at a number of different groups. As an investment vehicle, it was marketed at overseas buyers who would gain rental income and capital gains from letting to young professionals working in the new 'City of London' situated at Canary Wharf. The development was also targeted at young families and thus incentives for first-time buyers were offered. The development comprised 50 per cent buyers and 50 per cent tenants. These units were seen as comparatively cheap in the London housing market. A single bedroom flat costs £160 000 and a flat could be rented for just under £1000 a month (at 2003 prices).

Planning requirements (so-called 'section 106' obligations) obliged the developer to provide 30 per cent affordable housing within the scheme. Consequently, in the last phase of the development three blocks of social rented housing were to be provided let by three housing associations. This part of the estate was expected to be ready for occupancy in December 2003. The estate manager explained that differential access to estate facilities would apply and that tension between the different groups might follow from the opening of the social housing blocks:

> The residents of the housing association blocks will have access to some but not all of the developments facilities, [such as] the restaurant and coffee bar but not the gym or swimming pool ... there will be a view that the housing association blocks may not be a welcome feature of the estate for the private residents. (Interview)

However, problems soon appeared in relation to maintaining the standards of the estate and the blocks: litter, security doors left open by a large number of absentee landlords and the turnover on the estate of private renters.

Security was one of the features of the estate and this included: CCTV cameras linked to a reception area, a concierge which would eventually be staffed 24 hours a day, site security patrol night checks, an emergency mobile number for residents, and access point fob keys for all resident blocks and the car parks. In addition to these features, residents were offered extra day and night security cover (although there would be an additional charge for this service). Residents were encouraged to establish a neighbourhood watch scheme and the estate manager attended regular liaison meetings with the local police.

The development included 'temporary' gates while development work was in progress. However, these gates, which had a robust and sculptured quality, did not give the impression of being temporary. The estate manager confirmed us that residents were happy with the gated entrance. Residents had also assumed these gates were a permanent feature of the development. However, the planning agreement required these gates to be dismantled and retractable bollards were installed in their place in 2003.

To the casual visitor (and many residents) the estate looked like a gated community with patrolling security and gated access staffed by security guards. In fact, it was intended to be a development that would have no gates but would only limit the public right of way to walking access. The development could be an example of what Low (2003) has called a 'faux-gated' community.

The gates became a major issue on the estate because of criminal incidents within the neighbourhood. The estate manager, the planning officer and local residents all identified crime and fear of crime as key reasons why the residents wanted the gates to stay. In letters to the council planning department and at a meeting with the planning officer, residents claimed that if the gates were removed and the public access footpath through the estate was reopened more residents would become victims of crime. Officers stated that overseas property owners had been contacting the council because their tenants were advising them about how dangerous the area was and that the gates they thought permanent were in fact only building site gates. The planning officer and the estate manager reported that sales were decreasing and that rents had adjusted downwards as a consequence of these security concerns.

The planning officer stated that gated community developments were a new issue for the planning team. Gates were previously allowed in the borough but these situations were described as entirely different in that developments were situated on private land with no public access. However, in similar developments gates had been disallowed despite petitions from residents. The planning officer stated that with reference to the current development:

> The developers erected gates without planning permission. Obviously some sort of makeshift security gate was required as expensive building materials were present on the site. However, these gates had a 'permanent' feel from the start. (Interview)

The planning department agreed to retractable road bollards to control entry but an application would need to be submitted for the gates to become a permanent fixture. The request to gate a public parked area was refused and any replacement for the current temporary gates was thought likely to be vetoed.

The original planning brief stated that although there would be no provision for vehicular traffic, a public access route would be a feature of the development. Therefore keeping the gates in place was thought to be contrary to the spirit of the provisions in section 106. The council was keen to uphold this situation and any argument to the contrary, it was suggested, would have to be presented very convincingly. As discussed earlier, petitioning for gates goes against the current government advice on good urban design practice and mixed development guidelines. Additionally, the legal implications would need to be thoroughly assessed.

The planning officer advised that at a recent residents' association meeting the main concern was security, particularly "that the gates be a permanent feature as there have been a number of incidents ranging from vandalism to actual physical assault" (Interview). As one officer commented:

> Issues arose at the initial planning meeting for the scheme concerning the potential lack of integration into the wider ... community from prospective residents. Several of these buyers have subsequently called claiming that they thought the estate was more exclusive than it actually is, and saying that tenants now wish to vacate their flats as they fear for their safety. (Interview)

The second major issue was access to a public garden located on the edge of the estate. Residents wished this to remain private as they were paying a service charge for its upkeep and maintenance and therefore felt it was inappropriate for non-residents to use (and possibly abuse) it. Furthermore, residents were concerned that if the community was to be open-access the council would not pay the bill for any vandalism or graffiti that might occur. As one estate resident noted, it was a private development and the council would have no liability for any damage. The planning officer stated at the meeting she was "concerned with the residents' exclusive attitude" (Interview). In turn, the residents were frustrated by what they perceived as an unsympathetic response to their anxieties.

This example illustrated the conflict between the planning department's responsibilities to protect 'rights of way' and promote 'permeability' (ease of movement in an area) and the desire of the residents to secure a safe environment in which to live. These gates had become a focal point around the management of higher income housing in an area of acute deprivation, with a high level of crime and fear of crime. What the example shows is that the battle to maintain gating represented an important area of conflict between residents and council staff and between principles of safety and security on the one side and those of community, neighbourhood and social cohesion on the other. However, a conceptualisation of gating as a club good can challenge the necessity to think in strict dimensions of public and private, allowing discussion to proceed at a less emotive level.

Conclusions

Gated communities via their costs of ownership and membership partition the population into 'beneficiaries' and 'non-beneficiaries'. However, gated communities provide one example from a long tradition in clubbing together for increased individual benefits. Other examples include trade unions, friendly societies, squash and bridge clubs. All make a distinction between members and non-members to determine the allocation of benefits and costs.

The gated community represents a dilemma for policy makers between the concepts of segregation and security. If consumer choice points to the desirability of privacy and safety as priorities for residents, it becomes increasingly problematic for policy makers to deny the exercise of this choice. To suggest that residents be denied security merely based on an abstract notion of social cohesion could be construed as paternalism.

The view that the 'gated community increasingly seems a misnomer for a highly privatised mode of living' (Atkinson *et al.*, 2003) assumes a relationship between neighbourhood cohesion and community development, based on an idealised model of housing design. As society has become more fragmented and privacy is highly desired by residents, to see gating as the antithesis of social cohesion by reinforcing social and class divisions, producing new forms of segregation between rich and poor, ignores the much more complex relationships between individuals and their environments. The evidence from these two case studies suggests that whilst there is some validity in these arguments, they are too simplistic in capturing the complex choices that residents make in their attachment to urban neighbourhoods. Undoubtedly gated communities represent a choice to exclude others, but as a club good, they may also represent a more positive model of housing development.

In the case studies both the fear of crime and actual crime levels have either resulted in gates being erected or in the demand for temporary gates to be made permanent. The examples illustrate that developments help to reduce residential segregation in areas that otherwise would have accommodated either multi-deprived households exclusively or have been used for other purposes.

Research (Manzi & Smith-Bowers, 2004) on a housing estate with a large number of social landlords responsible for the communal services and facilities that tenants enjoyed, showed how ineffective local residents felt in influencing and getting a better service from their landlords. Institutions such as homeowners associations and Common Interest Housing Developments can provide useful models of self-managed, territorial organisation, in conjunction with other more traditional residents' associations. In one of the case studies the Home Owners Association had been able to secure the gating of the estate to reduce crime, to protect motor vehicles and to prevent unsolicited entry. Outside the gates, the consortium of landlords could offer no such service.

The theory of club goods illustrates an alternative model of conceptualising gated developments. By providing a hybrid model of property ownership and rights alongside a representation of new forms of territorial organisation, the theory can extend an understanding of the function of gated developments that provides a more detailed insight into this increasingly common phenomenon.

The process of collective ownership and management may serve to increase permeability as much as decrease it. The development of an active residents' association in both cases can provide an opportunity to develop links across tenure divides. The consequence may well be that such neighbourhoods are less segregated in socio-economic terms than would be the case if the gating were not available. By protecting property prices and offering opportunities for social mixing (albeit in limited terms) gating may present opportunities for urban renewal that are at present poorly understood.

Rather than attempting to prevent the spread of such developments, policy makers need to come to terms with the spread of gated communities in less emotive language. They should consider how issues of segregation can be balanced against the need to develop consumer choice and potentially increase social cohesion by providing new forms of sustainable communities, instead of railing against privatism, isolationism and particular interests.

Acknowledgements

The authors wish to thank Jordan Trimby who worked on the original research and two anonymous referees for their comments on an earlier version of this paper.

References

Atkinson, R., Blandy, S., Flint, J. & Lister, D. (2003) *Gated Communities in England: Final Report of the Gated Communities in England 'New Horizons' Project* (University of Glasgow and Sheffield Hallam University).

Blakely, E. & Snyder, M. (1997) *Fortress America* (Washington DC: Brookings Institution).

Blandy, S., Lister, D., Atkinson, R. & Flint, J. (2003) *Gated Communities: A Systematic Review of Research Evidence*, CNR Summary 12 (Sheffield Hallam University & University of Glasgow). Available at www.neighbourhoodcentre.org.uk.

Brunn, S., Williams, J. & Zeigler, D. (2003) *Cities of the World* (Latham: Rowman & Littlefield Publishers, Inc).

Buchanan, J. (1965) An economic theory of clubs, *Economica*, 32, February, pp. 1–14.

Caldeira, T. (2000) *City of Walls* (Berkeley, California: University of California Press).

Coleman, A. (1985) *Utopia on Trial: Vision and Reality in Planned Housing* (London: Hilary Shipman).

Davis, M. (1990) *City of Quartz: Excavating the Future in Los Angeles* (London: Verso).

Department of the Environment, Food and Rural Affairs (DEFRA) (2003) *Closing Alleyways in the Battle Against Crime* (London: DEFRA News Release 30 July) Available at www.defra.gov.uk/news/2003.

Department of the Environment, Transport and the Regions (1998) *Modernising Local Government: Local Democracy and Community Leadership* (London: The Stationery Office).

Department of the Environment, Transport and the Regions (2000) *Index of Deprivation* (London: The Stationery Office).

Garland, D. (2001) *The Culture of Control: Crime and Social Order in Contemporary Society* (Cambridge: Cambridge University Press).

Glassner, B. (1999) *The Culture of Fear: Why Americans are Afraid of the Wrong Things* (New York: Basic Books).

Glasze, G. (2003) Private neighbourhoods as club economic and shareholder democracies. Unpublished paper, Belgeo, Bruxelles.

Gottdiener, M. & Hutchision, R. (2000) *The New Urban Sociology*, 2nd edn (Boston: McGraw Hill).

Home Office (2001) *Community Cohesion: A Report of the Independent Review Team Chaired by Ted Cantle* (Norwich: The Stationery Office).

Jordan, B. (1996) *A Theory of Poverty and Social Exclusion* (Cambridge: Polity Press).

Landman, K. (2003) Alley-gating and neighbourhood gating: are they two sides of the same face? Paper presented at the conference Gated Communities: Building Social Division or Safer Communities?, Glasgow, 18–19 September.

Lindstrom, M. & Bartling, H. (Eds) (2003) *Suburban Sprawl* (Latham: Rowman & Littlefield Publishers, Inc).

Low, S. (2003) *Behind the Gates* (London: Routledge).

McKenzie, E. (1994) *Privatopia: Homeowner Associations and the Rise of Residential Private Government* (New Haven and London: Yale University Press).

McKenzie, E. (2003) Common interest housing in the communities of tomorrow, *Housing Policy Debate*, 14(1/2), pp. 203–234.

Manzi, T. & Smith-Bowers, W. (2004) So many managers, so little vision: registered social landlords and consortium schemes, *European Journal of Housing Policy*, 4(1), pp. 57–75.

Miller, N. & Brewer, M. (1984) *Groups in Contact: The Psychology of Desegregation* (Orlando, FL: Academic Press).

Mumford, K. & Power, A. (2002) *Boom or Abandonment: Resolving Housing Conflicts in Cities* (Coventry: Chartered Institute of Housing).
Newman, O. (1972) *Defensible Space: People and Design in the Violent City* (New York: Macmillan).
Office of the Deputy Prime Minister (2000) *Quality and Choice in Housing: A Decent Home for All: The Housing Green Paper* (Norwich: The Stationery Office).
Raco, M. (2003) Remaking place and securitising space: urban regeneration and the strategies, tactics and practices of policing in the UK, *Urban Studies*, 40, pp. 1869–1887.
Royal Institute of Chartered Surveyors (RICS) (2002) *Building Balanced Communities: the US and UK Compared*, Leading Edge report series (London: RICS).
Rykwert, J. (2002) *The Seduction of Place: The History and Future of the City* (New York: Vintage Books).
Samuelson, P. (1954) The pure theory of public expenditure, *Review of Economics and Statistics*, xxxvi, pp. 387–389.
Sanchez, T. & Lang, R. (2002) *Security Versus Status: The Two Worlds of Gated Communities*. Draft Census Note 02:02, November (Metropolitan Institute at Virginia Tech).
Sandercock, L. (2002) Difference, fear and habitus, in: J. Hillier & E. Rooksby (Eds) *Habitus: A Sense of Place*, pp. 203–218 (Aldershot: Ashgate).
Scott, A. J. (Ed.) (2002) *Global City-Regions Trends, Theory and Policy* (Oxford: Oxford University Press).
Smith, S. (1989) *The Politics of Race and Residence: Citizenship, Segregation and White Supremacy in Britain* (Cambridge: Polity Press).
Tiebout, C. (1956) A pure theory of local expenditure, *Journal of Political Economy*, 64, pp. 416–424.
Webster, C. (2001) Gated cities of tomorrow, *Town Planning Review*, 72, pp. 149–170.
Webster, C. (2002) Property rights and the public realm: gates, green belts and Gemeinschaft, *Environment and Planning*, 29, pp. 397–412.
Webster, C. & Wu, F. (2001) Coase, spatial pricing and self-organising cities, *Urban Studies*, 38, pp. 2037–2054.
Webster, C. & Wai-Chung Lai, L. (2003) *Property Rights, Planning and Markets* (Cheltenham: Edward Elgar).

INDEX